Organisation, Location and Behaviour

Organisation Location and Behaviour

Decision-making in Economic Geography

PETER TOYNE
Lecturer in Geography
University of Exeter

© Peter Toyne 1974

All rights reserved. No part of this publication
may be reproduced or transmitted in any form
or by any means without permission

First published 1974 by
THE MACMILLAN PRESS LTD
*London and Basingstoke
Associated companies in New York Dublin
Melbourne Johannesburg and Madras*

SBN 333 14355 8 (hard cover)
SBN 333 14422 8 (paper cover)

Printed in Great Britain by
REDWOOD BURN LIMITED
Trowbridge & Esher

The paperback edition of this book is sold subject to the condition that it shall
not, by way of trade or otherwise, be lent, re-sold, hired out, or otherwise
circulated without the publisher's prior consent in any form of binding or
cover other than that in which it is published and without a similar condition
including this condition being imposed on the subsequent purchaser.

For Angela

Contents

List of Figures   x

List of Tables   xiii

Acknowledgements   xv

1 LANDSCAPE ORGANISATION   1

  1.1 *Systems*   3
      1.1.1 Elements and attributes; 1.1.2 Relationships
  1.2 *Landscape as System*   6
      1.2.1 Delimitation; 1.2.2 Structure

2 THE DECISION-MAKER   17

  2.1 *Information*   18
      2.1.1 Diffusion; 2.1.2 Search; 2.1.3 Inaccuracy and distortion
  2.2 *Preference*   32
      2.2.1 Image; 2.2.2 Scaling; 2.2.3 Uncertainty
  2.3 *Motivation*   40
      2.3.1 Beliefs and values; 2.3.2 Achievement; 2.3.3 Aesthetics; 2.3.4 Satisfaction

3 SCALE   54

  3.1 *Internal Economies*   59
      3.1.1 Division of labour; 3.1.2 Substitution; 3.1.3 Disintegration; 3.1.4 Balance of processes
  3.2 *External Economies*   62
      3.2.1 Competition; 3.2.2 Cumulative causation

4 Land — 71

4.1 *Demand* — 73
4.1.1 Growth; 4.1.2 Obsolescence and decay; 4.1.3 Intensity
4.2 *Evaluation* — 94
4.2.1 Physical; 4.2.2 Economic

5 Labour — 108

5.1 *Demand* — 109
5.1.1 Direct changes; 5.1.2 Indirect changes; 5.1.3 Induced changes
5.2 *Productivity* — 118
5.2.1 Introduction of machinery; 5.2.2 Education and social services; 5.2.3 Industrial relations; 5.2.4 Substitution
5.3 *Cost* — 129
5.3.1 Demand; 5.3.2 Supply conditions
5.4 *Mobility* — 133
5.4.1 Information; 5.4.2 Evaluation

6 Capital — 143

6.1 *Savings* — 144
6.1.1 Income; 6.1.2 Stimuli
6.2 *Investment* — 150
6.2.1 The market mechanism; 6.2.2 Innovation

7 Transfer — 161

7.1 *Direct Costs* — 161
7.1.1 Distance; 7.1.2 Commodity characteristics
7.2 *Indirect Costs* — 169
7.2.1 Transit times; 7.2.2 Safety and reliability; 7.2.3 Accessibility, flexibility and adequacy
7.3 *Friction* — 172
7.3.1 Movement minimisation; 7.3.2 Bid rents; 7.3.3 Agglomeration

8 Demand and Supply — 184

8.1 *Price* — 184
8.1.1 Demand schedules; 8.1.2 Supply schedules; 8.1.3 Equilibrium
8.2 *Location* — 191
8.2.1 Spatial supply; 8.2.2 Spatial demand; 8.2.3 Spatial equilibrium

9 CONSTRAINTS AND INCENTIVES	210
9.1 *Scale*	212
9.1.1 Farm reorganisation; 9.1.2 Monopolies, cartels and trusts; 9.1.3 Decentralisation	
9.2 *Land*	220
9.2.1 Planning law; 9.2.2 Covenants in conveyances; 9.2.3 Standards	
9.3 *Labour*	230
9.3.1 Conditions of work; 9.3.2 Mobility	
9.4 *Capital*	234
9.4.1 Taxation; 9.4.2 Inflation; 9.4.3 Resource mobilisation; 9.4.4 Subsidies	
9.5 *Transfer*	241
9.5.1 Revaluation; 9.5.2 Policies	
9.6 *Demand and Supply*	250
9.6.1 Innovation; 9.6.2 Taxation and tariffs	
10 INTERDEPENDENCE	257
References	261
Index	279

List of Figures

1.1	System relationships	5
1.2	Hierarchical organisation	7
1.3	Superimposition of boundaries and percentage fields	8
1.4	Desire-line analysis	10
1.5	Nodal structure	11
1.6	Morphological system	14
1.7	Control system	15
2.1	Decision-makers	19
2.2	Contact frequencies	21
2.3	Derivation of information-probability field	22
2.4	Simulation of diffusion	24
2.5	Geographical perception	29
2.6	Time and the known environment	31
2.7	Indifference curves	37
2.8	Personality and risk	40
2.9	Spatial preferences	41
2.10	Sub-optimal land use	52
2.11	The decision-making system	53
3.1	Marginal returns	55
3.2	Services and scale	58
3.3	Incompatible production sizes	62
3.4	Spatial competition	63
3.5	Location of industrial activity	65
3.6	Retail outlets in the CBD	66
3.7	Cumulative causation (positive)	68
3.8	Cumulative causation (negative)	69
4.1	Land occupance	72
4.2	Income and economic development	74
4.3	Leisure demand	76

List of Figures

4.4	Growth of world population	79
4.5	Types of population growth	81
4.6	Population-density lapse rates	85
4.7	Correlation between businesses and size of settlements	92
4.8	Space required for services in urban areas	93
4.9	Spacing of high-rise development	93
4.10	Microclimatic variations	95
4.11	Ecological conditions and crop yields	97
4.12	Limitations on mechanisation	98
4.13	Landscapes of Scotland	100
4.14	Landscapes of SE England	101
4.15	Land-values lapse rates	103
4.16	Urban land-value surface	105
4.17	Land values in the CBD	106
5.1	Age–earnings profiles	123
5.2	Variations in absenteeism	125
5.3	Industrial stoppages	126
5.4	Labour demand	132
5.5	Migration lapse rates	136
5.6	Mobility and education	137
6.1	Consumption, savings and income	145
6.2	Capital flows in Spain	154
6.3	Productivity changes	155
6.4	Diffusion of services	157
6.5	Early and late adopters	158
7.1	Carriers' costs	163
7.2	Geographical structure of fares	166
7.3	Railway charges	167
7.4	Locational figures	174
7.5	Force diagrams	175
7.6	Bid rents	177
7.7	Land use	177
7.8	Urban structure models	179
8.1	Demand schedules	185
8.2	Supply schedules	188
8.3	Changes in supply and demand	190
8.4	Spatial supply and demand	192
8.5	Spatial equilibrium	195
8.6	Christaller's landscape	197
8.7	Löschian landscape	198
8.8	Löschian landscapes	199
8.9	Transport networks	200

8.10	Modified demand cone	201
8.11	Distance and consumer behaviour	202
8.12	Status and consumer behaviour	205
8.13	Distance, status and consumer behaviour	206
8.14	Sectoring	209
9.1	Land subdivided through inheritance	214
9.2	Assisted areas	217
9.3	Development plan (1947 type)	222
9.4	Linkage and development	242
9.5	Shrinkage of distance	244
9.6	Retail prices and travel costs	245
9.7	Grant-aided railways	248
9.8	Railway developments	249
9.9	Taxation and the supply curve	252
10.1	The system of landscape organisation	259

List of Tables

1.1	Flow matrix	11
2.1	Perception of shopping opportunities	32
2.2	Image criteria	33
2.3	Ranking of supermarkets	34
2.4	Retail imageability	36
2.5	Probabilities under different states of the system	37
2.6	Expected pay offs	38
2.7	Dichotomies proposed for evaluation of cultural values	43
2.8	Achievement and social class	48
3.1	Unit costs of production and scale of plant	57
4.1	Disposable incomes	75
4.2	Urbanisation of population	83
4.3	Physical blight	87
4.4	Amenity provision	88
4.5	Environmental blight	88
4.6	Crop yields	90
4.7	Land capability classification	99
5.1	Overtime working	111
5.2	Input-output matrix	113
5.3	Production coefficients (matrix X)	114
5.4	Inverse matrix B	116
5.5	Employment coefficients	116
5.6	Productivity increases	119
5.7	Rice yields and mechanisation	121
5.8	Production costs	128
5.9	Labour costs in retailing	129
5.10	Inter-industry and inter-regional wage rates	130
5.11	Labour turnover	136
5.12	Reasons for leaving employment	140

5.13	Staff attitudes	141
6.1	Net flow of financial resources to developing countries	151
7.1	Transport costs	162
7.2	General railway classification	168
7.3	Retail functions and ribbon development	181
9.1	Parker–Morris residential space standards	229
9.2	U.S. residential density guideline standards	229
9.3	Road and footpath widths	230
9.4	Recreational standards	230
9.5	Cost yardsticks	239
9.6	Tariff reductions	255

Acknowledgements

Any author is always in the perpetual debt of a large number of friends, acquaintances and colleagues. The present author is no exception, but he wishes to record particular thanks to Patricia Gregory, Clive Thomas, and Rodney Fry (the cartography staff of the Department of Geography at Exeter University) for drawing the diagrams; to Andrew Teed for his photographic expertise; to Barbara Whinham and Pauline Holmes for typing various parts of the final manuscript; to Exeter Cathedral and Tom and Jerry for spiritual and mental inspiration and, above all, to his wife Angela for correcting and typing the assembly of words and squiggles which comprised the several drafts of the manuscript, and for creating peace, quiet and gastronomic delights when they were most needed.

The author and publisher wish to thank the following for permission to reprint or modify copyright material: The American Academy of Political and Social Science, and the authors for a diagram in Harris and Ullman (1945) (figure 7.8B); The American Economic Association and the author for a diagram in Bain (1954) (table 3.2); The American Geographical Society, the editor of *Geographical Review* and the authors for two diagrams from Berry, Simmons and Tennant (1963) (figure 4.6C and D); The American Sociological Society for a diagram in Burgess (1927) (figure 7.8A); Edward Arnold Ltd and the author for a diagram in Haggett (1965) (figure 7.5); Constable & Co. Ltd, and the author for a table in Azzi (1958) (figure 4.11); The editor of the *Journal of Soil Science* and the authors for a table in Curtis, Doornkamp and Gregory (1965) (figure 4.12); The London School of Economics and Political Science and the authors for a diagram in Layard, Sargan, Ager and Jones (1971) (figure 5.1); Macmillan Education Ltd for three diagrams in Toyne and Newby (1971) (figures 2.5A, 4.16 and 8.5); The Macmillan Press and the authors for two tables in Medhurst and Parry-Lewis (1969) (tables 4.3 and 4.5); The Massachusetts Institute of Technology Press and the author for a diagram in Isard (1956) (figure 8.9); Princeton University Press and the authors for a diagram in Kemeny and Thompson (1957) (figure 2.8); The Regional Science Association of America and the author for a map in Lasuen (1962) (figure 6.2); The Regional Studies Association and the authors for two maps, one in Cracknell (1967) (figure 4.1) and the other in Fines (1968) (figure 4.14); The Royal Scottish Geographical Society for a map in Linton (1968) (figure 4.13); The Soil Survey of Britain and the authors for material contained in Bibby and Mackney (1969) (table 4.7); The University of Chicago Press and the authors for material in Berry, Tennant, Garner and Simmons (1963) (table 7.3), in Klukholn (1956) (table 2.7), and in Weber (1929) (figure 7.4); The University of Exeter for a table in Toyne (1971) (table 2.4).

The author and publisher also acknowledge the use of government copyright or non-copyright material from: *The Census of Distribution and Other Services* (H.M.S.O., 1966) (table 5.9); *The Census of Production* (H.M.S.O., 1958, 1963 and 1968) (tables 5.8 and 7.1); Department of Employment, *British Labour Statistical Yearbook* (1971) (table 5.10); *Department of Employment Gazette,* April 1972 (tables 5.1 and 5.11) and May 1972 (table 5.6); Department of Environment (Railway B division) unpublished data on grants for railway services (figure 9.7); *F.A.O. Production Yearbook* 1970 (table 4.6); Hoyt (1939) (figure 7.8C); Inukai (1971) (table 5.7).

1
Landscape Organisation

> 'Curiouser and curiouser!', cried Alice (she was so surprised, that for the moment she quite forgot how to speak good English).
> Lewis Carroll, *Alice's Adventures in Wonderland*

The reaction of an intruder from another planet on seeing the human landscape of this earth for the first time might well be very similar to that of Alice on realising that she was opening out into the 'largest telescope that ever was'. Concrete or redbrick jungles of people and their industries appear as aliens in the natural environment, which they are increasingly converting into an apparently formless, chaotic and often polluted or congested landscape system, in which problems of scarcity, growth and development occur in conjunction with problems of surplus, stagnation and decline.

Being capable of thought and ingenuity, man is able to modify and organise his natural surroundings to his own ends, but the environment often imposes certain limitations on this ability. Consider the theoretical possibility of a Robinson Crusoe landing on an island of varying topography in which natural resources are randomly distributed. In deciding where to settle, Crusoe may be faced with a number of alternative possibilities, though which one he chooses depends on his attitude and intellect. He may, for instance, try to assess as far as he can all the conditions of topography, site, vegetation, resource availability, drainage and defence, so that he can choose the alternative that appears to offer the 'best' of those conditions. On the other hand, he may simply make the choice on the basis of nothing other than intuition, whim or fancy. In either case the initial locational decision reflects the influence, whether positive or passive, of the environment. Sooner or later, a further decision must be made on how to produce enough food for survival; and once more the constraining influence of the natural environment may be felt if, for example, inappropriate crops are tried or if the same plots of land are used successively to produce one crop after another. The environment thus appears to establish certain 'outer limits' to Crusoe's actions, but within those limits there may exist a degree of freedom of choice. In a simple situation such as this, man is able to organise the landscape to his own ends subject to certain environmental limitations; provided those limitations are not exceeded, few problems of organisation arise.

But what of more complex real-world situations than this? Is there anything to suggest that man's relationship with the environment is any less compelling? It is true that, because of his innate inventiveness, man's technological ability to reduce the limitation of environment in a number of ways becomes greater as economic development proceeds. Yet it is equally clear that the process of applying technology to environmental conditions tends to create a new series of problems, many of which are difficult to remedy: overproduction, soil erosion and the pollution of atmosphere, land, sea and rivers, are all evidence of this tendency. At the same time, economic growth, which is closely associated with the development of technological skills, tends to create further environmental problems. This is because population growth and improvements in standards of living increase the demand for land and natural resources, both of which are in finite supply, at a time when the development of technology tends to reduce the demand for labour which is in ample and increasing supply. Resource imbalance then ensues. Social and economic problems of unemployment or short time working may have to be resolved, and land and resource exploitation and development may have to be controlled in order to create a fair and adequate distribution and intensity of their use. Conflicting agricultural, industrial, commercial, residential and recreational demands must be reconciled, and new ways found of increasing the effective 'yield' of land and resources; and both of these imperatives may lead to the creation of a further series of problems or even intensify the problems that they were originally designed to resolve. Residential development, for instance, may have to take place at some distance from industrial development because of the limited availability of land; and because land which is in short supply becomes costly, the density of development may have to be very great. High-rise apartment blocks on the edges of cities may be an inevitable outcome of the dilemma created by the characteristics of scarcity and growth, but they also tend to create various problems of day-to-day living. Commuting times and costs are increased, congestion is accentuated, and life in large blocks of apartments becomes impersonal: nervous disorders and illnesses increase and the general quality of life symptomatically declines. In many ways, therefore, man's organisation of the natural environment may become paradoxically (though understandably) alien and possibly even self-destructive.

An intricate mechanism thus characterises man's decision-making within the environmental framework, and this mechanism must be clearly appreciated and understood if future decisions are to be made with due regard to their full implications and repercussions, and if the resulting system of landscape organisation is to function efficiently and effectively.

1.1 SYSTEMS

Systems analysis appears to offer a productive conceptual framework for the study of complex structures and mechanisms. Biologists, botanists and zoologists, for instance, have gained a deeper insight into the functioning of biological, botanical and zoological phenomena by focusing attention on the intricate interrelationships of ecological systems (ecosystems); various social scientists have been able to develop a better understanding of political, social and economic phenomena by recognising and analysing the mechanisms, structures and functioning of political, social and economic systems. Indeed, so widespread has the recognition and study of systems in many disparate disciplines become that, in one sense, an essential 'unity of science' appears to be developing (Bertalanffy, 1951). Since the relationship between man and his environment appears to be characterised by mechanisms that are no less intricate than those of any other system, a systems approach should provide a useful framework for the analysis of spatial structure and organisation.

By definition, a system consists of 'a set of objects (or elements) linked together with relationships between the objects and between their attributes' (Hall and Fagen, 1956, p. 18). Such a definition could equally apply to the many regional systems and organisms whose description has been the basic theme of geographical studies since the days of Ritter, Herbertson and Roxby; and indeed it is perhaps because regional landscapes are characterised by so many different elements and relationships that 'the notion of a system is not in any way new to geographic thought' (Harvey, 1969, p. 467). It is, however, only relatively recently that greater emphasis has been placed on the concept of landscapes as systems, and a more rigorous and systematic analysis made of the mechanisms that give rise to their structure and performance.

1.1.1 *Elements and attributes*

The ELEMENTS (objects) are the basic units of the system, though their definition and identification are dependent on the scale at which the system is analysed. In systems terminology, the scale of analysis is often referred to as the RESOLUTION LEVEL of the system (Klir and Valach, 1967; Harvey, 1969). For example, in a hi-fi stereo system, the elements may be thought of as the loudspeakers, the amplifiers and the turntable deck. Yet, at a different level of examination, each of these elements in themselves can be considered as separate systems; the amplifying system is thus made up of a series of elements, which includes transistors, diodes and capacitors. Similarly, at another scale of analysis, each of the elements of the amplifying system can be regarded as a system in itself; in this way, the transistors may be considered as systems of which the emitters, collectors and the base contact constitute the elements. Within any system, therefore, each of the elements constitutes an effective subsystem whose identification is entirely dependent on the resolution level of the initial system.

Each of the elements of a given system possesses certain ATTRIBUTES (properties) and, because it is these attributes that give the elements their individuality, it has sometimes been suggested that the true elements of a system are the attributes of the elements and *not* the elements themselves (Kuhn, 1966). The attributes of the loudspeakers of the hi-fi system would include such qualities as size, resonance, reliability, age and condition, so that by this definition it would be these attributes and not the loudspeakers themselves which would form the true elements of the system. In a sense, this is merely logistic hair splitting, for any element can *only* be defined by its true attributes; in order, therefore, to recognise an element of the system, it is necessary to be able to define its attributes.

1.1.2 *Relationships*

Interdependence among the elements and their attributes is 'a diagnostic property of systems' (Chorley and Kennedy, 1971, p. 2). Certain relationships of interdependence are essentially non-functional in nature and help to determine morphological structure, whereas others give rise to the functioning and performance of the system. In the hi-fi system, for example, the relationships between the location, size, reliability, age and condition of the loudspeaker, turntable and amplifying systems are in no direct sense functional, but they do determine the nature and structure of the system as a whole: in other words, it is these relationships which establish the MORPHOLOGICAL SYSTEM. The various inputs and outputs of the loudspeakers, turntable and amplifiers are, however, relationships of a different kind, for it is these which give the system its *performance*. Such input and output relationships among the elements and their attributes constitute the FUNCTIONAL or CONTROL SYSTEM which, through its mechanisms, may give rise to change in the morphological system.

The actual form of the relationships—or, as they are occasionally called, the links—can be either direct or indirect, in series or parallel, and may be of a feedback nature. Patently, certain links between the elements are direct (figure 1.1A), while the influence of many others may be felt only in an indirect, or looped, way (figure 1.1B): for example, the turntable of the hi-fi system is not directly linked to the loudspeaker, but is indirectly related to it via the amplifier. A series relationship is the simplest form of direct link by which *one* element is related to *one* other element (figure 1.1A), whereas a parallel relation relates *one* element to *several* other elements directly and simultaneously (figure 1.1C). The feedback relation is perhaps the most

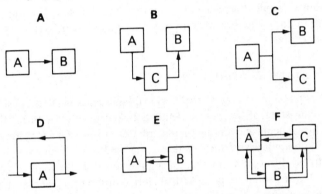

Figure 1.1 SYSTEM RELATIONSHIPS A—direct; B—looped; C—parallel; D—feedback; E—direct two-way; F—looped two-way.

intriguing of all because it describes the link, and consequently the mechanism, by which an element can affect itself (figure 1.1D). When a change is introduced into any one element of a system, there normally follows a series of repercussions throughout the whole system, the net effect of which may be that the initial change 'comes home to roost', as it were, with the element from which it originated. Such feedback can take several forms: it can be either direct (figure 1.1E), indirect, or looped (figure 1.1F), and positive or negative. Positive feedback occurs when the effect of an initial change in the system is snowballed to produce changes in the same direction throughout the whole system, and the result is that the initial change becomes self-perpetuating. Conversely, negative feedback has the effect of stabilising or reducing the effects of some original modification to the system so that an element of self-regulation, or *dynamic homeostasis,* is introduced. The crucial problem for system structure is whether its feedback relationships are largely of a negative or positive type. If they are largely negative, then dynamic homeostasis leads in the long run to a new state wherein any change in input is equated by a change in the structure itself; if they are largely positive, the initial change is self-perpetuating until it reaches the stage when the structure of the system can no longer cope with it, and the final result may well be the self-destruction of the system. The *strength* of the relationships may be just as crucial as their *nature* in determining the structure and performance of the system in which they occur. Some of the links in the system may be much stronger than others, and it may well be that just one strong but direct series relationship between two elements may have a greater total effect on the whole system than two or three weak relationships of positive feedback.

It is thus necessary to analyse the form and strength of all the various relationships among the elements and their attributes if the mechanisms, organisation and structure of morphological and control systems are to be fully understood.

1.2 LANDSCAPE AS SYSTEM

While it is possible to suggest that regional landscapes may be considered as systems consisting of a series of elements linked together with relationships among the elements and among their attributes, it is very difficult to establish the precise boundaries of such systems. The landscape of one area may well appear to be different from that of another area; and, as such, it may be thought that it should be possible to identify different regional systems. But this is the thorny old problem that has confronted geographers for many years because, in practice, it is notoriously difficult to define or identify regions adequately. Indeed, many would agree with Kimble that 'to spend our days regionalising is to chase a phantom and to be kept continually out of breath for our pains' (Kimble, 1951, p. 174). Yet it is clear that, if regional landscapes are to be considered as systems, they must be adequately defined and delimited (section 1.2.1) before their detailed analysis can begin.

1.2.1 *Delimitation*

By definition, a system is a coherently functioning unit whose constituent elements are functionally interrelated through the mechanisms of the control system. In the human landscape, a nodal region can be considered to be such a functioning system. Geographers have for a long time recognised that regions possessing some kind of functional coherence and internal interdependence of their constituent elements appear to characterise both physical and human landscapes (Robinson, 1953)—though the precise identification of such regions has been a difficult process which has been the subject of considerable debate.

One of the major problems of identification is that of resolution level. While it is clear that functional regions are characterised by the organisation of human activity around some nodal point in the landscape, it is equally certain that a hierarchy of nodal points can be recognised and that the organisation of human activity itself is essentially hierarchical in character (Garner, 1967; Haggett, 1965). Thus, nodal systems may be centred upon the home, the street, or the neighbourhood within villages or towns and cities of different hierarchical levels, each of which constitutes a subsystem within a larger system (figure 1.2; Philbrick, 1957). Similarly, various subsystems characterise the hierarchical structure of drainage basin systems in the physical landscape (Gregory and Walling, 1973). Which of the systems is chosen for analysis may depend on a number of factors related to the geographical scale of study which can be undertaken at any given time, so the resolution level of the system may also be determined by the scale of investigation.

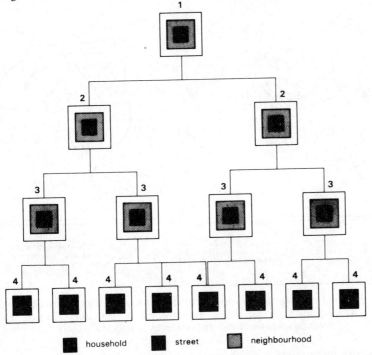

Figure 1.2 HIERARCHICAL ORGANISATION *Key* 1, conurbation; 2, city; 3, town; 4, village.

The delimitation of nodal regions at any resolution level is further complicated by the fact that their boundaries are elusive, and each of the many criteria that may be used to identify those boundaries gives rise to different geographical definitions of them. Consider the problem of trying to define a nodal region based on a medium-sized town. Many different activities take place in the town, each of which has its own hinterland (interaction field) which may be different in geographical extent from any other. If an accurate delimitation is to be made of the nodal region which the different activities together comprise, the interaction field of each activity must be established. As this would obviously be a physically impossible task, a representative sample of the town's various activities has to suffice—including industrial, retail, wholesale, educational and social activities. Each of the sample fields is then superimposed on a map. A pattern of interdigitating and meandering lines emerges, from which the maximum extent of the nodal region is easily identified (figure 1.3A). Similarly, it is clear that a 'core area'

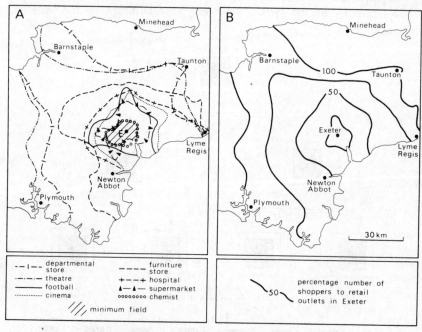

Figure 1.3 SUPERIMPOSITION OF BOUNDARIES AND PERCENTAGE FIELDS
A—superimposition of fields delimited by 8 indicators (Exeter, 1970);
B—percentage fields based on retail customers (Exeter, 1970).

may also be recognised which represents the region of ground common to all of the selected indicators (figure 1.3A). Between the maximum and minimum so defined there lies a zone within which interaction with the nodal centre varies considerably. It has been suggested that some intermediate boundary within these inner and outer limits may be taken to indicate some form of 'average' field. A useful concept is that of the 'mean' interaction field, which defines the area within which 50 per cent of the total movements in the maximum field occur; similar 'percentage fields' can be constructed for different values, for example, 30, 20 per cent and so on (figure 1.3B). Likewise, it has been suggested that a 'median' boundary line can be defined within the zone (Green, 1955), but it is not exactly clear whether any of these intermediate boundaries really represent anything meaningful (Haggett, 1965, p. 246). Indeed, the superimposition of boundaries appears to be useful only in a very general way, for it tends to create rather more problems than it solves: the maximum field clearly incorporates into the nodal region areas which have very weak associations with the nodal settlement, while the core area definition would exclude territory which is patently related to the node, and any intermediate line is merely arbitrary. We are, indeed, kept 'continually out of breath', as Kimble suggested, in attempting to hunt down the region by this means.

For this reason, other attempts at delimitation have been based on the use of only one indicator which is chosen as being characteristically representative of all other indicators. Such a method merely puts the proverbial cart before the proverbial horse! A further possibility lies in the analysis of the various forms of transport operative between the node and all the other settlements within its interaction field. Green (1950) and Carruthers (1957), for example, were able to define the hinterlands of the towns of England and Wales based on the frequency of bus services on market days. Godlund (1956) also used this method to delimit nodal regions in Sweden. The problem with this method is simply that bus-traffic flows show only bus movements, and it is by no means certain that other traffic movements within the region are of a similar nature. Detailed transport surveys of *all* movements would clearly give a more accurate picture of the real interaction pattern of the area. Yet taxi flows, pedestrian flows, car flows and rail flows all exhibit different patterns in just the same way as we find with the use of the different indicators for the 'superimposition' method. The same problem is confronted again: when more than one regional indicator is used, the difficulty of synthesis results.

The situation is complicated further by the existence of zones of overlap between the cores and the maximum extent of each nodal region. In theory at least, nodal structure could be identified by visual inspection of a map of desire lines based on actual flow data. On such maps, a line is drawn linking each settlement with every other place to which a flow exists, and then the boundary between one nodal region and another is interpolated fairly simply by assigning to each node all the settlements to which it is linked (figure 1.4A). In most real-world situations, a far more complex pattern of spatial linkage than this occurs; in such circumstances, visual inspection of the flow map reveals nothing save confusion (figure 1.4B). However, an alternative method of flow analysis—graph theory—does appear to offer some possibility of identifying nodal structure from flow data arranged in matrices.

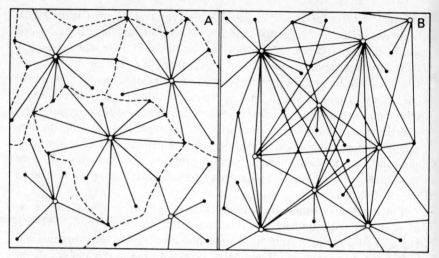

Figure 1.4 DESIRE-LINE ANALYSIS lines are drawn linking the pairs of places between which flows take place. A—a simple situation in which interpolation of boundaries (dotted lines) is possible; B—a more complex situation requiring more rigorous analysis.

Matrices are simply rectangular arrays of numbers arranged in columns and rows, any one entry of which is known as an element. Each column and row refers to a place in the study area, places a, b, c, d, etc. being listed in the same order along the rows as down the columns (table 1.1). The actual volume of flow between each pair of places is then entered in the appropriate cell of the matrix. For example, in table 1.1 there are 80 units of flow (buses per week) between Hornby and Trumpton. The largest flow from any one place is known as the *dominant flow* from that place. The total amount of flow *from* one place to all other places is known as the *out-degree* of the point and is indicated by the value of the sum of the elements in the appropriate row of the matrix. Similarly, the total flow *into* one place from all other places, known as the *in-degree* of the point, is indicated by the value of the sum of the elements in the appropriate column of the matrix. In order to identify the basic hierarchical structure of the flows, it is first necessary to rank all the points according to the value of their in-degree. Points from which the dominant flow is to a point of *lower* in-degree than themselves are the NODAL or TERMINAL points upon which the nodal structure of flows is centred, and it is to these points that the dominant flows of all other points are directed.

Table 1.1 shows a matrix of flow structure among ten hypothetical towns (matrix *X*). If *all* the flows were plotted as desire lines, a complicated structure would emerge from which it would be difficult to identify any

TABLE 1.1 FLOW MATRIX: bus connections per week

To From	Hornby	Trumpton	Nutwood	Fulton	Fordbridge	Axmouth	Stoneleigh	Morton	Summerton	Seaville	Total out-degree
Hornby		80	20	25	31	7	7	5	20	3	198
Trumpton	*76†	–	50	42	43	10	22	0	16	35	297
Nutwood	10	62†	–	12	52	3	2	7	9	14	171
Fulton	20	84†	2	–	41	8	31	42	22	12	262
Fordbridge	7	42	*51†	22	–	6	10	3	39	34	214
Axmouth	8	12	8	8	3	–	32†	6	2	5	84
Stoneleigh	3	18	7	9	16	*34†	–	12	22	7	128
Morton	4	0	7	2	3	7	1	–	12	46†	82
Summerton	20	31	46	2	9	3	62	4	–	72†	249
Seaville	12	22	4	9	31	4	12	7	*82†	–	183
Total in-degree	160	351	195	131	229	82	179	86	224	228	1868

† indicates dominant flow.
* indicates terminal point.
  see also figure 1.5.

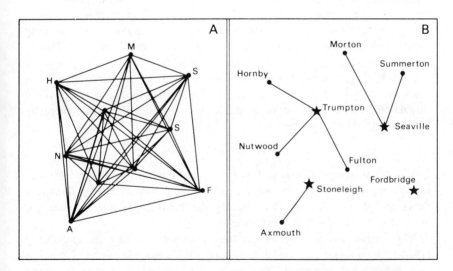

Figure 1.5 NODAL STRUCTURE A–desire lines between places indicated in table 1.1; B–nodal structure assessed by matrix analysis (nodal points indicated by the starred locations (Trumpton, Seaville, Stoneleigh and Fordbridge)).

nodal structure (figure 1.5A). However, using the method described above, Trumpton, Fordbridge, Stoneleigh and Seaville are identified as being the nodal points, and when all the dominant flows to these points are drawn in, a clearer picture (figure 1.5B) of the different regions of nodal interaction within the whole area is revealed.

No account has yet been made of any of the indirect associations which exist between pairs of settlements, yet these may be just as important as the dominant flows upon which the analysis has so far been exclusively based. The first step in accounting for indirect influence is to adjust the raw data matrix so that every element is expressed as a ratio of the *maximum in-degree* (column total) of the matrix: that is, by dividing every element of the original matrix by the maximum column total. Let this matrix be called matrix Y.

This kind of proportional matrix has several important properties. If it is squared or cubed (that is, raised to any given power), the values of each element of the resultant *power-series matrix* reflect all the possible indirect associations of a length equal to the power of the matrix, between the two places to which the element refers. In other words, the values of the elements of the Y^2 matrix account for all possible two-step indirect associations between pairs of places; the values of the elements of the Y^3 matrix account for all three-step associations; the Y^4 matrix accounts for all four-step associations; and so on. The final power-series matrix is the one that refers to the associations by the maximum number of indirect steps it is possible to make in the locational system being analysed. If each of the power series matrices is then summed, a final matrix (called the B matrix) results, in which each element represents the *total* direct and indirect associations occurring between all pairs of places (Haggett and Chorley, 1969; Nystuen and Dacey, 1961). The calculation of the B matrix would, of course, be a lengthy procedure if this method were followed for a large geographical area. A short-cut method based on the concept of the IDENTITY MATRIX (I) of the original matrix X is, however, available, the matrix B being calculated by means of the expression

$$B = (I - Y)^{-1}$$

The nodal structure of matrix B is then established by the same method as before: the nodal points are identified and then their associated dominant flows are plotted, but in this case the dominant flows now account for indirect as well as direct associations.

This technique was first successfully employed by Nystuen and Dacey (1961) to analyse the nodal structure of Washington State, based on long-distance telephone-call flows among forty cities in and near that state; six nodal systems were identified from the total system of flows. The latent nodal structure of Ireland has also been identified by graph theoretic methods based on postal flows among major settlements (Brook, 1973).

Other techniques have also been used to identify nodal regions from flow data. Goddard (1970), for example, has analysed the nodal structure of taxi flows in the City of London using factor analysis; while Garrison and Marble (1962) have used principal component analysis to identify the structure of internal airline routes in Venezuela.

Nevertheless, the problem remains that there is no completely adequate method of identifying the boundaries of nodal landscape systems with any real precision. In real-world situations, zones of transition are rather more typical than boundary lines, and most regions tend to merge imperceptibly with others, even though they may in certain respects appear to possess their own unique individuality of form or function. The definition of precise spatial limits to regional systems is perhaps rather more of an operational necessity than a recognition of geographical reality.

1.2.2 Structure

Once defined, the nodal region may be regarded as a system whose elements comprise the visual objects of human landscape occupance: the various residential, industrial, commercial, agricultural, transportational and recreational features which give every region its unique character of physical infrastructure. However, the recognition of elements in the landscape is dependent on the resolution level of the system. Hence, the residential elements may be considered variously as the individual dwelling units in their successive groupings into streets, neighbourhoods and districts within hamlets, villages, towns or cities. Similarly, the elements of the industrial landscape constitute a hierarchy of units ranging from individual factories through to industrial belts; the elements of the agricultural landscape range from individual farmsteads to agricultural belts; and the elements of the recreational landscape vary from individual parks and open spaces to large tracts of land of outstanding natural beauty.

At whatever resolution level they are considered, all these elements have certain attributes, such as their location, size, age and condition, which give them their individual character. It is the relationships among these various attributes of the residential, industrial, commercial, agricultural, transportational and recreational elements which constitute the morphological system of landscape organisation (figure 1.6; chapter 10).

The functioning of any morphological system is determined by the nature and strength of the various inputs and outputs among its elements. In the human landscape system, all the inputs and outputs emanate from the various decisions that individuals and groups make about the organisation and location of all forms of economic activity. The decision-making process thus constitutes the control system which gives the morphological structure of human landscape organisation its performance (figure 1.7).

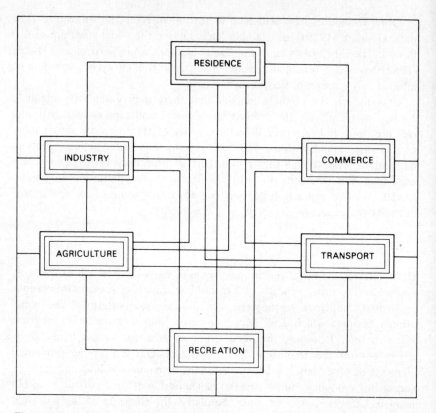

Figure 1.6 MORPHOLOGICAL SYSTEM direct and looped relationships link the residential, industrial, commercial, agricultural, transportational and recreational elements of the system. The different forms of each element which may be identified at various resolution levels are schematically indicated by the three self-contained boxes representing each element.

A wide range of physical, economic, political and socio-cultural elements enters into the decision-making process. Considerations of *scale* (chapter 3) are fundamentally important because they may condition not only the profitability of any enterprise but also its locational requirements. Similarly, the cost, availability and allocation of the factors of production—*land* (chapter 4), *labour* (chapter 5) and *capital* (chapter 6)—affect the possibilities of the organisation, location and profitability of human activity. Moreover, since distance separates all locations in geographical space, considerations of *transfer* (chapter 7) inevitably enter into the locational decision; and the conditions of *demand and supply* (chapter 8) are perhaps the major economic determinant of production and consumption levels within the system. All of these elements of the decision-making process are mutually interrelated and

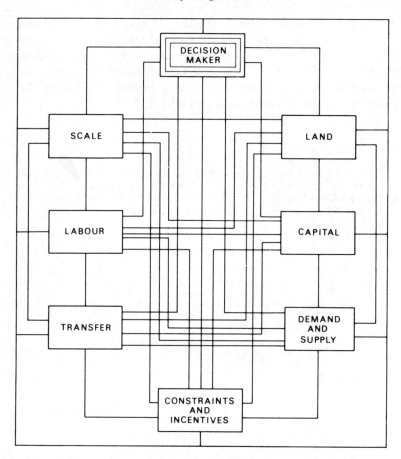

Figure 1.7 CONTROL SYSTEM direct and looped relationships between decision makers of various hierarchical levels and each of the elements of the decision environment.

affect each other to varying extents. Some of their relationships tend to be self-regulating, so that some of the mechanisms of the control system are essentially of a negative-feedback nature; but others tend to be rather more self-perpetuating, and thereby give rise to mechanisms of a positive-feedback nature. Decision makers may therefore find it necessary to introduce a series of *constraints and incentives* to their own decision-making process (chapter 9) in order to counteract the mechanisms of the control system that they themselves have created. The evaluation of each of these elements is conditioned by an equally wide range of elements relating to the information, preferences and motivation of different decision makers, who consequently may be viewed as a 'subsystem within a socio-cultural system defined by a wide range of economic, political, physical and social units and relationships' (Golant, 1971, p. 204; figure 2.11). The spatial control system is a complex mechanism based on direct, looped, series, parallel and feedback relationships between the attributes of decision makers and each of the various elements of their decision environment. It is upon the mechanisms of the system that the spatial organisation of human landscapes is based.

FURTHER READING

Ackerman (1963); Bunge (1962); Curry (1964); Curry (1966); Eyre (1964); Hartshorne (1939); Herbertson (1905); Jackson (1972); McLoughlin (1969); Morgan and Moss (1965); O'Riordan (1969); Roxby (1926); Smailes (1971); Stoddart (1963); Tatham (1951); Vita-Finzi (1969).

2
The Decision-Maker

Decision-making is a far from haphazard process. Consider the problem of deciding where to go for a belated holiday sometime after Christmas. First of all, as much *information* as possible has to be collected about alternative places, costs and dates. The various travel agents' brochures will undoubtedly provide a considerable amount of factual (or semi-factual!) data which may form a starting point for making the decision; but a number of other sources, including television programmes and hearsay from friends, relatives or colleagues, may also be used to give supplementary detail. The extent to which such information is readily available will, in large measure, influence the decision at the outset. In places where complete information about the alternatives is available, any individual should be able to make a better decision than he would be able to make in places where only partial knowledge can be obtained. Additionally, the way in which the information is searched and learnt may also affect the decision being made. After reading three or four of the travel brochures, the intending holidaymaker may well decide either that he has not enough time to go through all the other brochures or that he has already found one place at the right price that will do. The order of search may well, therefore, be important in determining the outcome of the decision. Having searched the available information, there may well be several alternative propositions that seem attractive, and in order to decide which one of these to choose, the list has to be organised into a series of *preferences*. Such preferences are determined by the individual's assessment of the expected utility of each of the alternatives, which in turn is related to the individual's *motivation*. The assessment of alternative holidays, for instance, may be based on such criteria as: which appears to be best value for money; which seems most likely to live up to expectations; or which entails the least risk. But the final choice may well be conditioned by the extent to which the individual really wants to find the very best, or whether he is simply concerned with finding a holiday that 'will do'.

Decision-making is thus based on a relationship between information availability and the preferences and motivation of the individual decision maker. But there are many different kinds of decision maker, ranging from individuals to whole groups and organisations (figure 2.1), each of which behaves as a system at a given resolution level, and which may be differently motivated and express a different set of preferences based on different information.

Once a decision is made affecting any one of the elements, a series of repercussions may well be felt throughout the whole set of elements in the system. The extent of these repercussions is dependent on the size of the alteration made to the first element, and also on the *strength* of the relationships among the elements and their attributes (section 1.2). Some of the effects of the initial decision may be of limited importance to other parts of the system, but more often, especially where the relationships are positive and strong, the effects are felt throughout the system. These consequences may represent the desired effect of the initial decision, but more frequently it is found that the effects of a particular decision have either not been far reaching enough or else have been too great. Such is the constant dilemma that faces planners dealing with problems of landscape organisation and economists dealing with problems of unemployment and inflation. Decision-making is consequently a cyclical process because of the feedback nature of the relationships among the elements of the landscape system.

2.1 INFORMATION

Information about the decision environment is generally obtained both indirectly by the inevitable diffusion of certain information through any given society, and also directly by the search of the decision maker for a particular fact or series of facts. However, rarely (if ever) does any individual manage to gain complete or perfect information about any given situation. It is because of this limited knowledge that decisions are normally less than optimal.

2.1.1 *Diffusion*

Information is transmitted through a population by two slightly different methods that may be distinguished on the basis of the number of people involved. Where several people receive a piece of information from *one* other person without its being altered by some third party (or parties), that information is called PUBLIC INFORMATION. It is normally transmitted either by 'direct spoken communication to an audience, by radio and television, by postal circulars, by publication or by film' (Hägerstrand, 1967,

The Decision-Maker

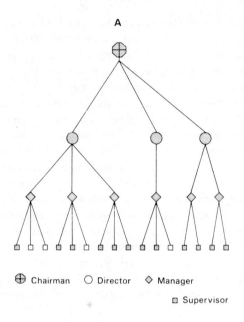

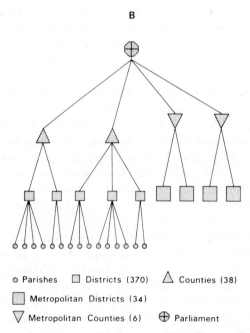

Figure 2.1 DECISION MAKERS A—hierarchical structure of management; B—hierarchical organisation of government decision making (England, 1973).

p. 139). On the other hand, information that is passed from person to person, either directly by word of mouth or indirectly by letter or telephone, is normally known as PRIVATE INFORMATION.

In either case, the speed with which information can be disseminated depends very largely on the individual's frequency of contact with the appropriate source. Contact-frequencies are not very easy to ascertain, but there is certainly a close relationship between distance and the number of contacts made by an individual. Hägerstrand (1953) has shown that in Sweden, telephone contacts decrease rapidly with distance from any given telephone exchange (figure 2.2A-B): a pattern that appears to be characteristic of many other areas, and is similar to the relationship characterising letter postings and visits made by individuals to friends and relatives (figure 2.2C). Contact-frequencies are consequently fairly circumscribed in their geographical extent, and this fact underlies not only the way in which information may be diffused but also many aspects of human behaviour. The distances over which people migrate (Hägerstrand, 1953) and the areas from which they choose their marriage partners (Perry, 1969; figure 2.2D) are typical reflections of the 'neighbourhood effect' of contact-frequencies (section 5.4).

Other factors are, of course, contributory to the rate of information diffusion. Patently, the actual distribution and density of population in an area may directly influence the possible contact-frequencies of any given individual; in a similar way the actual physical relief of an area may present certain obstacles to settlement or to communication, thereby producing the same effect.

Various attempts have been made to demonstrate theoretically the way in which the spread of information is affected by these contact-frequencies (Hägerstrand, 1953; Yuill, 1965; Gould, 1969; Cliff, 1968). Other factors (such as relief and distribution of population) are, of course, important, but in order to test the effects of contact-frequencies *alone*, certain simplifying assumptions have to be made about those other factors in rather the same way that the assumption of *ceteris paribus* is often made in economic theory. Thus, the simplifying assumption that is initially made about the environment through which information is diffused is that it is characterised by a flat, featureless plain in which the population is uniformly distributed. A further simplifying assumption is that transport is taken to be available in all directions at a cost increasing in direct proportion to the distance involved. Such an environment, or land surface, is known as an ISOTROPIC SURFACE. This surface can be graphically represented as a matrix of squares, each containing a given number of people. The matrix shown in figure 2.3A represents such an isotropic surface divided into cells, each 1 kilometre square and containing say, ten persons, with the initial carrier of the information living in the centre of the area.

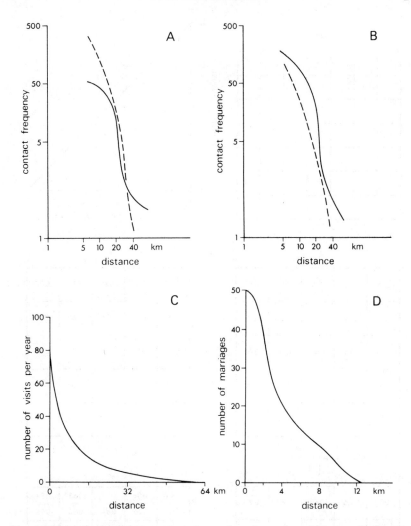

Figure 2.2 CONTACT FREQUENCIES A—number of outgoing telephone calls (Exeter—dotted line; Foix [Ariège, France] —continuous line); B—number of incoming telephone calls (Exeter and Foix); C—visits to friends and relatives (sample of population in Thurcroft, Yorkshire); D—number of marriages (church weddings, Thurcroft 1962-8).

Within this environment contact-frequencies will, as in reality, decrease with distance from the carrier of information. Let it be supposed that the distance–contact relationship takes the form shown in figure 2.3B. From this graph it is possible to construct a matrix showing the number of contacts that may be made in each of the squares of the actual isotropic surface. It can be seen that the information carrier in the centre of the area will come into

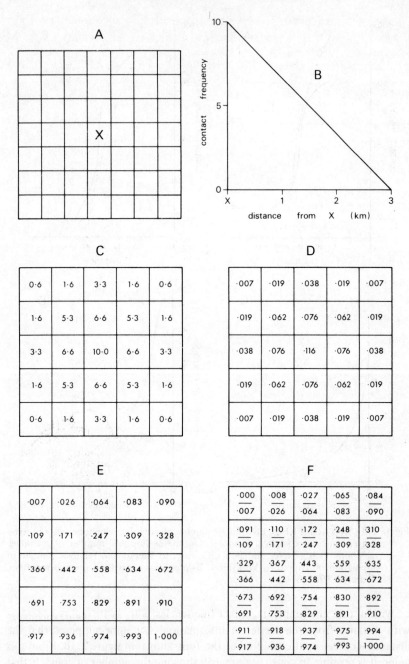

Figure 2.3 DERIVATION OF INFORMATION-PROBABILITY FIELD

contact with ten persons in any given time period: so the number 10 is entered in the centre cell of the matrix that is now being constructed (figure 2.3C). The appropriate contact-frequencies may then be entered for each surrounding cell. When completed, this matrix is processed to produce an information *probability field*. The 'processing' follows three simple steps.

(1) The number in each cell is divided by the sum total of all the numbers in the grid (in this case, the total is 86.0). The resulting grid, which is usually called the MEAN INFORMATION FIELD (MIF), is shown in figure 2.3D. It will be observed that the sum of the resulting values now equals 1.00.

(2) Beginning with the cell at the top left-hand side and progressing along each successive *row* of the MIF grid, the values in each cell are added cumulatively to produce the ACCUMULATIVE INFORMATION FIELD (AIF) (figure 2.3E). That is to say, the value of cell 2, row 1 (.019) is added to the value of the preceding cell (.007) to produce a cumulative value of .026 for cell 2, row 1.

(3) Each cell is then ascribed a *range* of values whose lower limit is prescribed by the value in the cell preceding it (following the same sequence used in step 2), and whose upper limit is set by its own value (as given in the AIF matrix). The new matrix is known as INFORMATION PROBABILITY FIELD (IPF); see figure 2.3F. It is so termed because it shows the relative probability of information being passed on from the carrier in the centre of the area. It will be noted that the range of values of the central cell is the greatest of the whole matrix, and that, as distance from this central cell increases, the range of values in successive cells decreases. This *range* of values in fact indicates the relative probability of the information being transmitted to a given area. Areas with a large range of values thus stand a greater chance of receiving information than do areas with a small range of values.

The process of simulating the diffusion of information over the whole isotropic surface is based on the IPF which shows the *likelihood* of the information being passed on to a particular person in a given area. As has been shown, the greatest chance is that the information will be passed on to someone in close proximity to the carrier. But which person is, in fact, chosen to receive the information? In reality, this is likely to be something that will be determined largely by chance, for it will probably be the person whom the carrier happens to see at a particular moment. This, of course, is just as chancy as drawing a given number out of a range of numbers from 0 to 100. Now, of course, the IPF in this context is nothing other than a range of numbers from 0 to 1, and it is in this context that its usefulness will become apparent. If the person to be contacted by the information carrier is, in reality, selected by chance, as has been suggested, then in the simulation of this possibility a number in the range 0–1 can be randomly drawn to

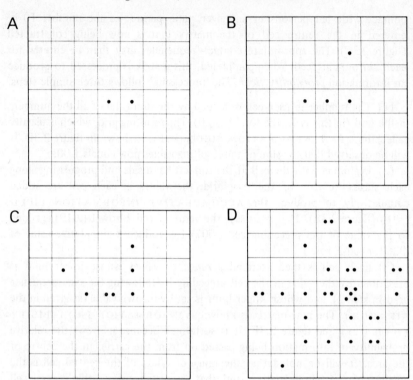

Figure 2.4 SIMULATION OF DIFFUSION each dot in diagrams A–G represents one

represent this chance. Since the IPF has greater *ranges* in the cells nearer the carrier, it follows that there will still be a greater chance that the selected contact will be found somewhere near the centre than at some distance away.

The simulation thus begins with the IPF being superimposed on the central cell (representing the initial carrier of information) of the isotropic surface. Conversations then take place at constant time intervals called 'generations', and in each generation every carrier passes on the message to one other person. So, in the first generation the initial carrier passes on the message to one other person who, in turn, becomes a carrier in the second generation. The person to whom the message is to be transmitted is determined by the random number selected from the range 0–1. (Random numbers can be selected from appropriate random-number tables contained in most elementary statistical tables, such as the *Cambridge Elementary Statistical Tables*.) Suppose the first number to be drawn is 0.572; this means that the contact person is found in the fourth cell of the third row across in the IPF matrix (that is, the cell next to the right of the initial carrier of the

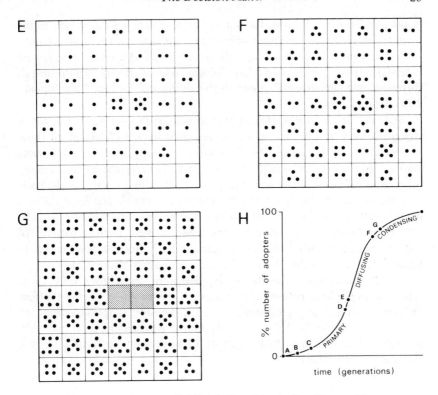

contact: maximum contacts per cell (10) is indicated by shading (diagram G).

isotropic surface); see figure 2.4A. In the next generation, the IPF is superimposed successively on the two cells which now contain the information carriers, and a random number is drawn respectively for each of them. The process is repeated through as many generations as are necessary for everyone on the surface to have received the message.

If the number of persons who have received the message at the end of each generation is plotted successively on a graph, a logistic, S-shaped curve is formed. Clearly, this curve describes the fact that, at first, the message is spread very slowly, but as more and more people become carriers the rate increases until the time is reached when only a few people have not yet received the information. Inevitably, it may take some time for these few to receive the message, so that towards the end the 'acceptance rate' slows down (figure 2.4H). The acceptance rate varies spatially as the information is gradually diffused further and further over the isotropic surface. In the initial stages (figure 2.4B–C) most of the acceptors are found near to the original centre of the diffusion. Later on, mainly during the period of rapid diffusion in the middle of the S-curve, the number of acceptors is smallest near to the original centre, becoming bigger with increased distance away from it (figure 2.4D–E), while in the later stages all areas increase at a similar rate (figure 2.4F–G). These three phases are known respectively as the 'primary', 'diffusing' and 'condensing' stages in the diffusional process (Hägerstrand, 1953; figure 2.4H).

Naturally, in the real world the process will hardly ever be as clear cut or straightforward as this, for it will be modified by the effects of all those factors which, in this theory, for the sake of simplicity, have so far been held constant (on the assumption of an isotropic surface). Hägerstrand and many other workers have shown how the patterns of diffusion vary as the basic assumptions are gradually relaxed to incorporate realistic and local patterns of population distribution and various barriers to communication (Cliff, 1968). Likewise, account may also be made of more realistic patterns of contact-frequency, not based on a simple straight-line relationship (as in figure 2.3B) but on reality (as described, for example, in figure 2.2).

Yet all these improvements in the design of the theory only cause variations of the basic pattern. Information, as theory shows, does not diffuse uniformly throughout a geographical area, neither does it spread at constant rates. Consequently, the location of any individual decision maker may well, in itself, affect that individual's store of information and help to determine the quality of his decision.

2.1.2 *Search*

However, few decision makers are likely to be entirely dependent on information that eventually comes their way through the gradual processes of diffusion. Most information is specifically sought either from various published sources or from other individuals or groups of individuals, but the extent to which this searching process is carried out, and the amount of information that is acquired by searching, are both fundamentally limited by the decision maker's availability of time or money (or both).

Time is a significant factor because decision makers are either often restricted to a time limit within which they must make their particular decision or else they may simply tire of the search and give up after a while. Equally, the availability of money may limit the search merely because, in certain circumstances, relevant information can only be obtained at a price (whether a direct price charged for its purchase or an indirect price representing the cost of postage, transport, telephone calls, expense-account lunches, industrial spy teams or any other expenses incurred in its acquisition). The availability of time and money may well in this way affect a decision maker's search for information and, where time and money are limited, the search may well be concluded just as soon as some *satisfactory* information is achieved, rather than when the *best* has been acquired.

Just how long it takes to find the first satisfactory information is very largely determined by the order in which the search is made. Consider, for example, the way in which a newly-engaged couple might set about the task of finding their first home. Information about all the possible houses which fall within their price range, size and locational requirements, can be sought from the local estate agents, who day by day send details of all the different houses that come on the market. The couple begin to look around the various possibilities; it may just happen that they strike it lucky and in the first day's postbag there are details of a house that seems to suit them reasonably well. The problem then faces them of deciding whether to snap up this desirable property before someone else does, or whether to wait a little longer to see if something better comes onto the market and run the risk of the earlier property being sold in the meantime. After all, who knows whether the next day's postbag will bring details of a more desirable 'chez nous'! On the other hand, it might be that the couple seem to be waiting for weeks on end for anything that looks even remotely acceptable to turn up and each day they get progressively more and more depressed, disillusioned and despondent until in the end they simply *have* to choose a house either because the date of the wedding has been fixed or because life is becoming so miserable that *any* house will be better than the eternal waiting. In either of these two situations it can be seen that the order and length of search is essentially determined by the randomness of events in the housing market in relation to the timing of the search and the time that is available for that search to be pursued.

The order of search may also affect the length of the search and the selection of a course of action by its indirect 'conditioning' of the searcher. Most of us have, at some time or other, made the comment that 'I still haven't seen anything as good as that one I saw some time ago', (whether it be in searching for bargains in the autumn sales, looking for the 'best buy' in washing powders, or whatever), and thereupon decided to stop searching any further and go back to the perceived 'best' alternative. In such circumstances, it may be the apparent excellence of an 'early' alternative that makes most of the successive alternatives appear so much worse and which may lead to the early termination of the search for a better alternative. Equally, a longish run of 'poor' alternatives from the beginning may make a 'better' alternative seem more attractive than would have been the case had it been found earlier in the search.

In this context, there appears to be something of a difference between individuals and larger groups of decision makers (such as large firms or corporations); it is often argued, for instance, that searching by individuals is rather less systematic and less exhaustive than that of large firms (Abler, Adams and Gould, 1971). One of the main reasons for this would seem to be the relative difference in the availability of time and money to the two kinds of decision maker, and this in turn, may well be a reflection of their differing motivations. Equally, there appear to be marked differences between individuals in their search behaviour and it seems likely that variations in individuals' age, status, personality, educational level and mental ability may underlie the extent to which they are prepared to make extensive or systematic searches for information (Day, 1969).

2.1.3 *Inaccuracy and distortion*

Both from the search process and by the processes of diffusion, individuals gradually build up a store of information relative to their decision making but it is rare, however, that the information so acquired is either totally accurate or entirely complete.

There is, for instance, ample empirical evidence to suggest that much of the information that an individual may have learnt about his geographical environment through successive searching and experience is often rather inaccurate. The relationship between actual location and an individual's perception of actual location is, in this respect, significant. In one survey in Exeter, for example, a sample of people who had lived in the city for at least one year were asked to assess the distance from a point in the centre of the High Street to certain well-defined landmarks (Toyne and Newby, 1971, p. 164). In general, it was found that people tended to underestimate distances to the eastern end of the High Street and to overestimate in the other direction. Furthermore, these 'guestimates' became progressively more inaccurate for locations at greater distances away from the city centre (figure 2.5A). It would also appear that individuals' mental maps of a given known area vary according to the location of the individual: the 'northerner's' view of Britain, for example, is vastly different from that of either the Welsh or Londoners (figure 2.5B-D) and none are remarkable for their accuracy!

Of course, such faulty and incomplete perception of geographical (and other) information is related not only to search behaviour but also to many other characteristics of the individual perceiver, who ultimately makes decisions based on his own faulty perceptions. The age, sex, income level, status, personality, educational level and mental ability of the decision maker have all been shown to influence his perception (Huff, 1960), and it would appear that, other things being equal, older people of higher income and education levels, and with extrovert personalities, tend to be rather more accurate in their perception than younger individuals with relatively low incomes and poor educational background.

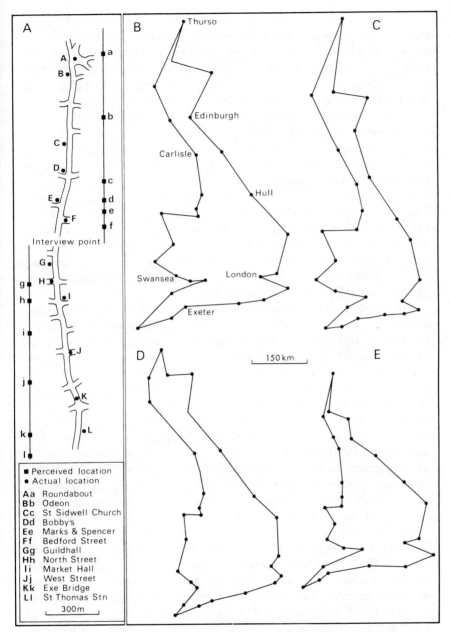

Figure 2.5 GEOGRAPHICAL PERCEPTION A–comparison of actual and perceived locations (Exeter CBD); B–actual location of 27 towns in Britain; C–perceived location of the towns shown in diagram B (sample of Welshmen); D–perceived location (sample of northerners); E–perceived location (sample of Londoners).

A major source of inaccuracy in individuals' information fields arises from the process of diffusion where 'private' information (defined above) is involved. It is notorious that messages passed by word of mouth can get distorted or added to inadvertently or deliberately. The almost legendary story of the message sent by the troop commander to brigade H.Q. (allegedly in World War I) is a useful reminder of the extent to which message distortion can occur. The commander sent the message along the line to H.Q.: 'Send reinforcements we are going to advance', but it finally arrived at H.Q. as: 'Send three-and-fourpence we are going to a dance'! In the real world, message distortion takes place every day to some extent, and even an individual's mental maps may incorporate such distortions, as they tend to be based, at least in part, on what other people have led the individual to believe. Thus, for example, on arrival in a new town a newcomer might be told by a well-intentioned neighbour that a well-known local beauty spot is several miles away, only to find later that in reality it is a good deal nearer than he had been led to believe. In the period before the spot was visited, however, the newcomer's behaviour would probably have been partly conditioned by the inaccurate information that had come his way.

The extent to which individuals acquire only partial information is, as has already been suggested, related to the search process and the time available. The mental maps that individuals possess are thus modified through time as the environment is gradually searched and learned (figure 2.6). Consider, for instance, how a housewife's mental map of the location of supermarkets is influenced by the search process: newly arrived in a particular town, she sets off to discover where the supermarkets are. The first one she finds will be the one nearest to the bus stop or car park that she happened to use. But 'nearest' does not necessarily mean the one which is, in *reality,* the nearest to the bus stop or car park, as much as the one that is nearest in the direction in which she happened to walk. If her time is limited she will probably have to use this particular store for her purchases on this particular trip. Next time, if unsatisfied with the supermarket she found on the first trip, she would probably go off in the opposite direction, hopefully to find somewhere better. She *might* be lucky! If she is not, she will repeat this random search for some while until eventually she is satisfied. There may well, however, remain in the (as yet), *terra incognita* of supermarketland an alternative that she may, alas, never discover. On the other hand, she might have been lucky and found just what she wanted at the first attempt, in which case her *terra incognita* would be as large as that marked on some of the earliest world maps! It may, of course, just be possible that the newcomer was a lady of very systematic method and unlimited wealth—in which case, *terra incognita* would not, for her, exist. Such an individual is, however, rarely found!

More normally, as a result not only of random searching through

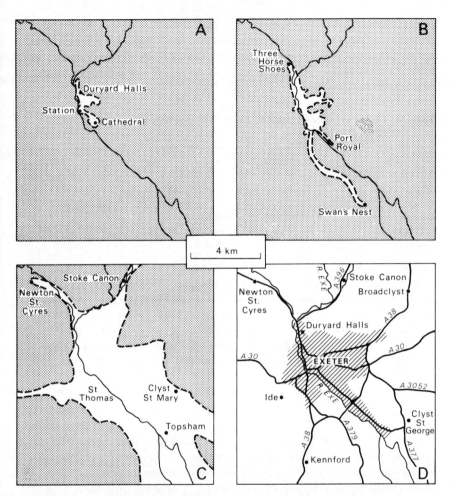

Figure 2.6 TIME AND THE KNOWN ENVIRONMENT students' perception of Exeter and surrounding district. Shaded areas are not known by 75 per cent of the sampled students. A—extent of area 'known' by undergraduates living at Duryard by the end of their fifth week in Exeter; B—area known by same students by the end of 30 weeks in Exeter; C—area known by the end of 3 years; D—the actual area under consideration. (The significance of visits to public houses in the learning process is, by inference, apparent!)

geographical space but also of the random way in which information is diffused to them, most people eventually build up a picture of geographical information. That this information is, to say the least, partial, can be seen in the results of a survey of shoppers in Exeter who were asked to list the shops that they felt might offer the goods for which they were looking (Toyne, 1971). The difference between the perceived and actual opportunities varies considerably but, in general, it is clear that customers have a very limited perception of the real alternatives available to them. Thus, only a half or even less of the actual number of greengrocers, bakers or chemists are known to most shoppers, while practically all shoppers know the location of every supermarket in town (table 2.1).

TABLE 2.1 PERCEPTION OF SHOPPING OPPORTUNITIES: City of Exeter, 1971. Source: Toyne (1971).

Shop Type	Mean number perceived for every 10 shops of given type	Shop Type	Mean number perceived for every 10 shops of given type
Grocer	6	Furniture	8
Greengrocer	5	Ironmonger	7
Butcher	6	Bookseller	7
Fishmonger	6	Jeweller	5
Baker	5	Hairdresser	4
Clothing	9	Chemist	4
Shoes	8	Supermarket	10

The effect on the decision-making process of such distortion and inaccuracy of both diffused and searched information is, of course, considerable for it leads to essentially sub-optimal behaviour in the expression of preference ratings.

2.2 PREFERENCE

Fundamental to the decision-making process is the assumption that the decision maker can evaluate in some comparative way all of the alternative courses of action that may be available to him. Generally speaking, he does this by assessing the *utility* of the various outcomes of the available courses of action, and then on that basis expresses his preferences among the alternatives offered. In this way, the evaluation of preference is very closely related to the decision maker's information field and the way in which he has perceived, sought and learnt about his decision environment.

2.2.1 *Image*

In order to express a preference, the decision maker must be able to list a number of criteria by which to measure the relative merits of the alternative courses of action available. These criteria are responsible for creating in the decision maker's mind the 'image' of the particular item being considered.

Take, for example, the criteria by which alternative retail stores may be evaluated. Undoubtedly, one consideration must be that of the prices charged for similar commodities in the various alternative shops. This may take the form of a simple comparison of whether a particular store's prices are in general low, high or fair in relation to its competitors'; or it may be that a more quantitative assessment can be made. Similarly, comparison of the quality and range of goods on offer would, in all probability, be made. In fact, as Kunkel and Berry (1959) have indicated, a whole series of such criteria affect the retail image, and these range from assessment of price through to store atmosphere (table 2.2A).

TABLE 2.2 IMAGE CRITERIA: A—retail store image, based on Kunkel and Berry (1959); B—housing image, based on Ratcliff (1949).

A *Retail images*	B *Home buying images*
(1) Price of merchandise	Price of house
(2) Quality of merchandise	Rateable value
(3) Assortment of merchandise	Freehold/leasehold/rented
(4) Sales personnel (attitudes, numbers knowledgability)	Number of bedrooms Is there a separate dining room?
(5) Locational convenience with respect (a) to shoppers' home or place of work (b) to location of competitors	Size of kitchen Size of garage Construction—brick, stone, cob and thatch
(6) Other convenience factors such as availability of parking space hours open store layout	Additional liabilities (e.g. road charges) Provision of mains services (is there water, gas, electricity, drainage?)
(7) Services provided (credit, delivery, lifts, escalators, restaurants, pram parks)	Central heating Size of garden
(8) Sales promotion (are there frequent special events, lead-loss lines)	Accessibility relative to place of work
(9) Style, quality, reliability of advertising	city centre main roads schools
(10) Store atmosphere	social amenities
(11) Reputation for fair trading on adjustments or purchases	Social environment Privacy/density of development Future developments nearby

Likewise, an individual searching for a new home will assess the alternatives on criteria related both to the structural characteristics of the houses or flats and their location (Ratcliff, 1949). Thus, the 'imageability' of housing is based on such considerations as price, rateable value, number of bedrooms or reception rooms, garage space, privacy and location with respect to schools, workplace or entertainment facilities (table 2.2B).

In the industrial world there is evidence to show that entrepreneurs evaluate alternative sites in terms of transport, labour and linkage imageability (Cameron and Clark, 1966; Newby, 1971). Transport, or movement, considerations are included in most evaluative criteria and are, in themselves, subject to evaluation in terms of such images as cost and time (Huff, 1960).

The 'image' is thus based on a large number of criteria, all of which are liable to some fundamental inaccuracy not only in their evaluation but also in their scope, because of man's inaccurate and incomplete perception of the relevant information.

2.2.2 Scaling

Each of the available alternatives is next evaluated on each of the 'image' criteria, and there are three main ways in which such 'preference scaling' may be effected:

ORDINAL SCALING simply involves putting the alternatives into rank order of preference on each of the criteria in turn, and then adding up the rank scores. The alternative with the *lowest* total rank score should then be chosen as the most preferred. No units of scale are involved in this method, which is based on mere subjectivity, and results in a list of the form $O_1 > O_2 > O_3 > O_4$ (where O_x = outcome number x, and $>$ means 'is preferred to'). An example will clarify the method. Suppose that four supermarkets (A, B, C and D) are being evaluated according to the image criteria listed in table 2.2A. In terms of image 1 (prices) A is preferred to C, which itself is preferred to B, which in turn is preferred to D—that is, $A > C > B > D$—so that A scores 1 (rank 1), C scores 2 (rank 2), B scores 3 and D scores 4. The process is repeated for the remaining eleven images and then the ranks are summed. It can be seen (table 2.3) that, taking all twelve items into consideration and *assuming that each of the twelve items counts equally,* supermarket B will be preferred to D, A and C respectively—that is, $B > D > A > C$.

TABLE 2.3 RANKING OF SUPERMARKETS: criteria 1–11 as shown in table 2.2.

Supermarket	Criteria												Total	Rank
	1	2	3	4	5a	5b	6	7	8	9	10	11		
A	1	2	3	4	3	3	4	1	3	4	4	3	35	3
B	3	1	4	1	2	1	2	3	1	2	1	1	22	1
C	2	4	2	2	4	4	3	2	4	3	2	4	36	4
D	4	3	1	3	1	2	1	4	2	1	3	2	27	2

SCALINAL SCALING is a rather more refined technique that enables the decision maker to indicate the *strength* of his preference directly in such terms as 'I prefer A twice as much as B' (which could in this case be written $A > 2B$, and which therefore takes the general form $A > kB$ where k is a constant indicating the strength of the preference order).

TRANSINAL SCALING is a yet more refined method that can be used when the decision maker is able to specify *absolute* preferences in terms of some quantitative measuring rod. The problem, however, is that it is usually very difficult to find a suitable measuring standard that can be applied to the often diverse set of criteria on which preferences are being rated. Normally some monetary standard is applied and everything is scaled in terms of cost, value or profit. The enquiry into alternative sites for London's third airport attempted, in a classic cost–benefit analysis, to scale such diverse criteria as noise, landscape amenity and accident-liabilities in terms of their supposed costs (Roskill, 1971). There is, however, considerable doubt as to the validity of this method since many criteria may not be properly equated with monetary values. Alternatively, a simple 'point score' method can be used by which the alternatives are awarded a score in a given range of numbers (for example, 1–10) to indicate the extent of the individual's preference for them.

Despite the obvious basic simplicity of all three methods, there remains a major problem in synthesising all of the separately scaled criteria to produce a final preference rating. It may be necessary to do this because of the likelihood of INTRANSITIVE PREFERENCES being expressed where several image criteria have been used. Normally, if $A > B$ and $B > C$ it should follow that $A > C$, but it is often found that, in certain circumstances, C is preferred to A simply because one of the evaluative criteria is being weighted more heavily than the others (May, 1954). In evaluating preferences among three supermarkets, A might be preferred to B in terms of its range of goods, B preferred to C in terms of its personnel, yet C preferred to A because it gives trading stamps. In such a case of intransitive preference, it is necessary to try to equate the three criteria in terms of some common yardstick. It may well be that an individual decision maker may count one of his image criteria as being rather more important than another, so that it may become necessary to scale the image criteria themselves and 'weight' them according to some further scale. It can be shown, for example, that the criteria which affect retail image are ranked in different order according to different types of store (Toyne, 1971). In the context of supermarket imageability, the range and price of goods are considered to be more important criteria than the locational convenience, quality or service, whereas locational convenience is the prime consideration in the image of tobacco stores (table 2.4).

A further complication arises in that the decision maker may not always be able to rank all the alternatives completely. It may nevertheless be

TABLE 2.4 RETAIL IMAGEABILITY: ranking of criteria by shop type, City of Exeter, 1971. Source: Toyne (1971).

Shop type	\multicolumn{5}{c}{Criteria}				
	convenient place	prices	better selection	quality	service
Supermarket	3	2	1	4	5
Grocer	5	4	1	3	2
Greengrocer	5	3	2	1	4
Butcher	4	3	5	1	2
Fishmonger	2=	5	1	2=	4
Baker	4	5	2	1	3
Clothing	5	2	1	4	3
Shoes	5	2	1	4	3
Furniture	5	2	3	4	1
Ironmonger	3	5	2	4	1
Tobacconists	1	5	3	4	2
Hairdresser	2	3	–	–	1
Chemist	3	4	2	–	1

possible, in such a case, to discover the pairs of alternatives between which he is *indifferent*. Suppose that an individual has to choose between drinking sherry or beer in various combinations of quantity. It may be possible for him only to say that the same utility (measured in terms of such image criteria as happiness, drowsiness or whatever!) could be obtained from, say, four sherries and one beer as from two sherries and three beers, and that between these alternatives he was totally indifferent. Such indifferences can be plotted cartographically to produce INDIFFERENCE CURVES (figure 2.7). A whole series of indifferences can similarly be mapped to reveal, perhaps more clearly, where the real preferences lie.

2.2.3 *Uncertainty*

Many decision situations involve an assessment of either risk or uncertainty that may, in turn, affect the actual preference scaling among various alternatives. In theory, at least, the process of risk assessment may be simply described in mathematical terms of probability.

Suppose that it is necessary to find the best way of getting from Exeter to Bristol by road. A number of alternative routes (say five) may be suggested $(R_1 \ldots R_5)$, along which traffic conditions are known to vary. Let it be assumed that three sets of road conditions (*states* of the system) may exist: in the first set the road may be congested (S_1), in the second traffic may be very light (S_2), and in the third there may be an 'average' condition (S_3). Patently, travel times will vary according to route and conditions, and these can be presented as a matrix (table 2.5). An examination of this matrix may help one to decide which route to take. Travel times on route 1, for example, vary very considerably according to conditions, whereas a journey on route 5,

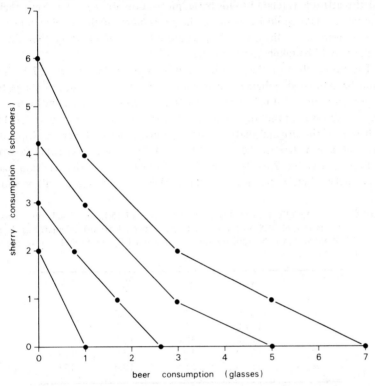

Figure 2.7 INDIFFERENCE CURVES hypothetical indifference preferences between various combinations of sherry and beer.

TABLE 2.5 PROBABILITIES UNDER DIFFERENT STATES OF THE SYSTEM: journey times in minutes between Exeter and Bristol by 5 alternative routes.

Routes	States of Congestion		
	S_1	S_2	S_3
R_1	220	90	125
R_2	195	140	180
R_3	210	100	125
R_4	220	110	140
R_5	200	105	130

albeit shorter under congested conditions than route 1, takes considerably longer than route 1 under 'light' conditions. Should the prospective traveller set out on route 1 and pray for 'light' traffic conditions, or what? Clearly, at this point, the individual's attitude to the risks involved will be all-important, and this attitude is related to his basic 'motivation' or character. Nevertheless, if some estimate could be made of the *probability* of there being congested, light or average conditions *at the particular time of the journey* then the risk element could be taken into account.

The probability of the various conditions occurring at, say, 17.15 hours could be estimated as being (for example) 60 per cent chance of congestion, 30 per cent chance of light traffic and 10 per cent of average conditions. This can be expressed in the form of a PROBABILITY VECTOR (0.6, 0.3, 0.1). If each row of the original matrix of conditions (table 2.5) were multiplied by this vector, a matrix of the EXPECTED PAY-OFF PROBABILITIES would be produced (table 2.6). It is on this information that the decision maker could make a better choice among the available alternatives, as will be shown.

TABLE 2.6 EXPECTED PAY-OFFS: journey times at 17.15 hours. Road conditions and routes as in table 2.5. Column E indicates expected pay-off calculated as the sum of all elements of the appropriate row multiplied by the probability vector (0.6, 0.3, 0.1).

	S_1	S_2	S_3	E
R_1	132.0	27.0	12.5	171.5
R_2	117.0	42.0	18.0	177.0
R_3	126.0	30.0	12.5	168.5
R_4	132.0	33.0	14.0	179.0
R_5	120.0	31.5	13.0	174.5

Now, in reality of course, as has already been emphasised, it would be rare for any individual to possess such complete and accurate information about conditions and probabilities. On the other hand, there is little doubt that individuals do set about the process of assessing risky situations by a method which, while not so rigorous, is at least similar to this mathematical method. Actual risks and probabilities are mainly replaced by perceived and estimated risks and probabilities known as SUBJECTIVE PROBABILITIES that are, again, a function of personality and experience.

The difference between actual and perceived risks is not easy to assess, although—as Kates was able to show in the context of storm hazard perception in the United States—most people react to risk by 'making events knowable, finding order where none exists, identifying cycles on the sketchiest of knowledge or folk insight and, in general, they strive to reduce the uncertainty of the threat of hazard. Or conversely, they deny all knowability, resign themselves to the uniqueness of natural phenomena, throw up their hands in impotent despair and assign their fates to a higher power' (Kates, 1967, p. 67). In the former case, their estimates of the probability of risk will very likely bear little relation to reality, yet this estimate is treated very much as a reality and is included in the decision process. Thus, farmers' perception of drought in the Great Plains of the United States is shown to diverge quite considerably from reality (there being up to 20 per cent difference between the farmers' estimates and the actual number of years in which drought occurred), yet there can be no denying that farmers bear this estimate in mind when deciding on their farm organisation (Saarinen, 1966).

Likewise, the decision maker's own attitude to risk may well be important in determining which of the several alternative situations or actions he is likely to adopt. Individuals react very differently from each other in risky situations. One person may be extremely cautious by temperament while another may even delight in embarking on a risky course of action. But the size of potential gain may well be an important factor in determining the decision maker's attitude; provided that the pay-off is high enough, even a moderately cautious person may occasionally take a reckless gamble.

Of course, it is not so much the known pay-off that is important as the perceived pay-off; and it is not without significance that there is a close relationship between the character of individuals and the way in which they perceive these pay-offs. Kemeny and Thompson (1957) have demonstrated the form of this relationship for different kinds of people. The reckless gambler, for instance, imagines all possible gains or winnings to be far larger than they are in reality, whereas the cautious person tends to underestimate their size (figure 2.8).

These factors help to explain why it is that, given identical and fully documented sets of alternative actions, different individuals will choose differently between them. Given the choice of which route to take between two locations, the reckless individual would probably take the one that, although involving an element of risk, *may* get him there in the fastest time; whereas the cautious individual would take the one involving no risk. In the situation described earlier (tables 2.5 and 2.6), the reckless gambler would probably choose route 1 and the extremely cautious driver would probably select route 2; yet it is route 3 that offers the minimum risk combined with maximum speed (the minimax solution).

The process of establishing preferences among alternatives thus involves the decision maker in an assessment of image, scaling and uncertainty; and it is his perception of these three characteristics that is so fundamentally important in determining his final choice. Consequently, the expression of

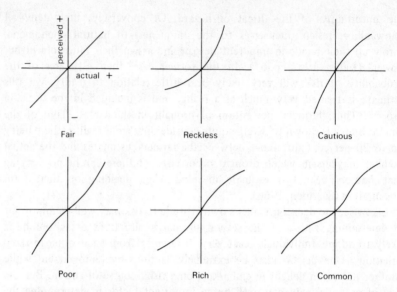

Figure 2.8 PERSONALITY AND RISK Source: Kemeny and Thompson (1957).

preference is also related to the factors affecting the decision maker's perception, so that individuals' spatial preferences vary quite markedly not only with their age, education and income level, but also according to their own geographical location (figure 2.9). This fact reflects, once more, the tendency for individuals to possess a limited and distorted range of information (Gould 1966; Gould and White, 1968).

2.3 MOTIVATION

Preferences are ultimately reflections of the decision maker's basic motivation. A number of characteristics appear to be significant in shaping the motivational patterns of individuals or groups, but many of these characteristics are themselves partly determined by motivation. Initially, motivation is conditioned by the beliefs of decision makers, which in turn may affect their desires to achieve something in life and their attitudes towards satisfaction. From these conditions arise their social and political values, and even their sense of the aesthetic, which may well help to determine certain of their actions. Motivation thus constitutes a complex subsystem of interrelationships, frequently involving feedback among the beliefs, desires, attitudes and values of the decision maker, all of which are ultimately related to his age, education and social background. The precise nature of this subsystem not only varies from individual to individual but also, importantly, according to the resolution level of the system. Groups, for instance, may be

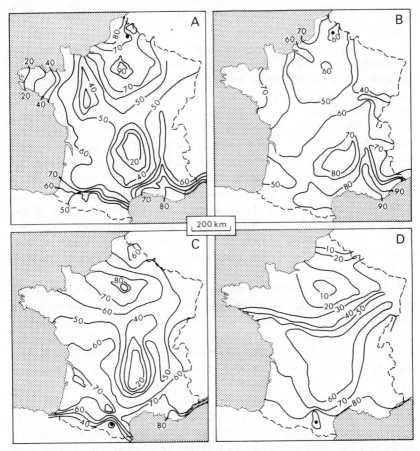

Figure 2.9 SPATIAL PREFERENCES differences in the attitudes of young and old Frenchmen in Lille (Nord) and Foix (Ariège) towards places where they would prefer to live; A—young people in Lille; B—old people in Lille; C—young people in Foix; D—old people in Foix. Notice, particularly, the preference of young people for the Paris region and the comparative aversion of older people to that region. (Contour line values indicate the percentage of the sample who would prefer to live at different locations.)

differently motivated from individuals—a fact that may well have important consequences in the decision-making processes.

2.3.1 *Beliefs and values*

The most rudimentary form of the intellectual expression of belief is that of myth: 'a primitive philosophy, the simplest presentational form of thought, a series of attempts to understand the world, to explain life and death, fate and nature, god and cults' (Bethe, 1905). Progressively, however, through increased rationality of thought, myth and fantasy have been gradually superseded, and a more rational theology has evolved in many societies. This movement from *mythos* to *logos* has normally been dependent on the emergence of a professionally trained priestly group of leaders who have sought to define a rational ethic based on the implications and traditions of their beliefs. In one form, rational theology is centred round a belief in a transcendent God: as for instance in Judaeo-Christian theology, where Yahweh becomes the focus of worship and belief. But rationalisation can also lead to the replacement of belief in a transcendent God by an entirely secular attitude to man and the world; in such a case the 'secularisation of culture' and society, with an emphasis on 'trust in self and man', results.

In either case, a 'moral theology' evolves and becomes a part of the established doctrine of the group, so that the members of the group acquire a set of moral or ethical standards that are inextricably 'bound up with practical attitudes towards the most varied aspects of daily life' (Parsons, 1958).

In this way social values may well be conditioned by the individual's or the group's belief patterns, even to the extent that belief becomes instrumental in the maintenance of social structure (Durkheim, 1954). The same also may be true, to a lesser extent, of political values and political structures, or economic values and economic structures. Certainly it is the case that 'in societies with or without a high level of technological or economic development, there is a definite economic system in which the religious (belief) system helps to motivate, guide, distribute and validate the productive and distributive energies of the individuals in it' (Goode, 1951, p. 137). Indeed, it is possible to go even further and suggest that both the nature and the success or failure of economic systems are closely related to belief. Weber, for instance, maintained that the emergence of a capitalistic economic system was largely made possible by the emergence of Protestantism, with its ethics of frugality, hard work and emphasis on saving (Weber, 1904).

Conversely, it may also be the case the belief is, in itself, partly conditioned by social or political values. It would be difficult, for instance, to suggest any other rational explanation of humanism and other forms of secular belief. In this situation, the 'norms' of society become embodied in its creeds, which in turn are instrumental in establishing its accepted norms, and the perpetuation of both creed and social values is thereby assured.

It follows that an individual's social values will be a reflection of his deep-rooted beliefs or philosophy of life. As such, they may incorporate his standards of *fairness*, his attitudes of *concern* and his *scruples*; all of which may 'guide the action choices of the individual by delineating the range of outcomes which he perceives to be acceptable in view of the other participants' (Chapin, 1965, p. 33). Such values vary considerably not only among individuals but also among different hierarchical groups of individuals, such as families, local government authorities and national governments. But, in general, it would appear that this sense of social values varies according to a number of 'dichotomies' (Parsons, 1951; Kluckholn, 1956; table 2.7).

TABLE 2.7 DICHOTOMIES PROPOSED FOR EVALUATION OF CULTURAL VALUES. Sources: Items 1–5, Parsons (1951); items 6–18, Kluckholn (1956).

(1) AFFECTIVITY–AFFECTIVE NEUTRALITY (whether motivated by immediate gratification or not).
(2) SELF-ORIENTATION–COLLECTIVITY ORIENTATION (whether motivated by self-gratification or not).
(3) UNIVERSALISM–PARTICULARISM (whether basing judgements on general standards or on particular circumstances).
(4) ASCRIPTION–ACHIEVEMENT (whether assessing others on the basis of their personality or achievements).
(5) SPECIFICITY–DIFFUSENESS (whether acting towards others in certain well-defined contexts only).
(6) DETERMINATE–INDETERMINATE (whether believing in order or chance in natural events and society).
(7) UNITARY–PLURALISTIC (whether the world is seen as a whole or as a series of different dichotomies).
(8) EVIL–GOOD (whether human nature is regarded as evil or good).
(9) INDIVIDUAL–GROUP (as Parson's self-orientation–collectivity orientation).
(10) SELF–OTHER (whether always concerned for self above others).
(11) AUTONOMY–DEPENDENCE (whether acting independently).
(12) ACTIVE–ACCEPTANT (whether accepting fate or trying to change it).
(13) DISCIPLINE–FULFILMENT (whether controlling self or promoting self-realisation).
(14) PHYSICAL–MENTAL (whether intellectual).
(15) TENSE–RELAXED.
(16) NOW–THEN (whether emphasis is placed on present or future events).
(17) QUALITY–QUANTITY.
(18) UNIQUE–GENERAL (whether events and phenomena are regarded as unique or recurring).

To a large extent it could be argued that present-day society is broadly based on 'affective neutrality, collectivity-orientation, universalism, achievement and specificity' (Gross, 1966, p. 68)—though individuals, as distinct from groups, are probably rather more orientated toward affectivity, self-orientation, particularism, ascription and specificity. Whereas, for instance, the planning department of a city may decide to shelve a scheme that has all the cost benefits of cheapness, in the interests of a small minority group of individuals for whom the plan would involve a degree of upheaval, it is rather more doubtful if an individual would put others first in such a way. Social values thus have a different character at different resolution levels of the system, and this may also be true of political values.

The acquisition of social and political attitudes and affiliations may be considered as a system in which information is distributed along various channels of communication to a number of nodal points. In such a system, the channels of communication are very diverse and diffuse, for they consist of all the possible means of information distribution available, such as person-to-person visits, attendance at meetings and all forms of literature and the mass-media generally. Individuals and all groups of individuals at various hierarchical levels (for example, families, households or neighbourhood groups) comprise the nodes of the system and may be thought of as senders, processors and receivers of information (Cox, 1969).

The spread of politically relevant information, for instance, is related not only to the nature of the available means of communication but also to the characteristics of the senders, processors and receivers. To a certain extent, therefore, political information is diffused in much the same way as any other, and the basic characteristic of information flow as being determined largely by distance also applies (section 2.1). But the spatial process of information diffusion may be modified by other characteristics than mere distance; in particular, the existence of different 'acquaintance circles' of various kinds affects the degree of 'chance' in contact frequencies.

Individuals have a marked tendency to select as friends and information sources those who hold similar attitudes to themselves, though there still remains considerable debate as to which comes first: the attitude or the friend. Homo-political selectivity of this kind has close affinities with the 'chicken and egg' syndrome in the old problem of cause and effect. It can be argued, on the one hand, that friends are selected once an individual has acquired his political views; while, on the other hand, it can be argued that it is through meeting various friends and by contact with mass media of many different kinds that political opinions are formed. Thus, it has been shown in Britain that 'individuals who read newspapers with an editorial policy favouring the Labour party are more likely to vote Labour than those who read Conservative newspapers, though there is clearly a problem of reader selectivity here in that individuals read papers which confirm previous attitudes' (Cox, 1969, p. 88).

Whatever the cause, the effect of homo-political selectivity is manifest in many different situations, and leads to such 'reciprocity situations' (Rapoport, 1957) as husbands and wives or parents and children influencing each others' political views to such an extent that they are usually almost identical. Similar processes are also typical of the acquisition of social values.

Political attitudes are also inextricably associated with the social and geographical structure of population in any given area. Where there is a strong and closely-knit social structure, there tends to be a certain homogeneity in the political attitudes of the population, and this tends to reflect the attitudes of the leaders of the community. In the Midlands of Britain, for example, it is in those parliamentary divisions where social structure has been centred on the clergyman and the local patron of the 'living' (usually a Conservative landowner) that there is a markedly uniform Conservative vote; in contrast, in Wales, where social organisation has been largely centred on the chapel, and where landowners have been treated with no special respect, there is a more homogeneous Labour vote (Cox, 1967). Similarly, there appears to be a rural–urban difference in terms of political homogeneity. Residential and social segregation and heterogeneity is thus believed to underlie the greater partisan division, which is evident in the political views of urban populations compared with the more homogeneous views of rural populations whose social structure remains equally homogeneous and strong (Segal and Meyer, 1968; Cox, 1968).

The process of 'political socialisation' (Hyman, 1959), by which individuals acquire their social and political values, is closely related to the personal characteristics of the individuals (or nodes) themselves. As might be expected, social class, educational background, personality and psychological attitude are all significant in this context. 'In every political system, the better-educated, higher social–economic status groups are more active in political organisation, more apt to vote, more apt to participate in any way in the political system' (Sorauf, 1965, p. 41). Likewise, there is a tendency for homo-political selectivity to correlate with age and educational level; in general, such selectivity is most evident in older people and in those with a poor educational background (Himmelstrand, 1960). Psychological attitudes have also been tentatively equated with political values. An individual whose personality traits include fear, suspicion and inability to cope with uncertainty or anything complicated is thus thought to seek leadership and authority from others, and tends to adopt a 'follow my leader' attitude that may act as a complement to any social structuring and dominance which may exist (Adorno, 1950).

However acquired, political and social values are significant in that it is from them that public policies normally emerge. Laws, planning decisions, treaties and all kinds of 'agreements' are normally created by political individuals or institutions (including pressure groups) acting on behalf of the individual in society (chapter 9). For the politician, the central problem is one of making sure that the objectives he pursues and the decisions he takes are truly reflective of the will of the individual or group(s) he represents, for it is only in this way that the beliefs, political and social values and aspirations of a society can be translated into reality.

2.3.2 Achievement

The extent to which individuals desire to achieve something in life is one of the main factors influencing motivation. It is notoriously difficult to assess what is going on in a person's mind, but psychologists have devised several techniques that attempt to measure a person's 'achievement motive'. These normally involve the collection of a sample of a person's thoughts, usually by getting subjects to write short essays prompted by a series of photographs, and then analysing the stories for evidence of 'achievement imagery' (ideas or fantasies that are expressions of the desire to 'get on' or achieve). In this way, 'a simple count of the number of such achievement-related ideas in stories written under normal testing conditions could be taken to represent the strength of a man's concern with achievement' (McClelland, 1961, p. 43). This count has been called the 'n-achievement' score of the individual. On this basis, there is plenty of evidence to show that men do have 'limited horizons' that reflect the extent of their achievement desires and that vary not only from individual to individual but also, noticeably, from region to region and from country to country.

The reasons for these differences in achievement desires are varied and complex, but initially at least seem to be related to the level and type of information or background knowledge that has been accumulated by the individual, group or country. Information, as has been shown earlier (section 2.1), can be either public or private in nature, and both sources may be important in creating different levels of achievement motivation in individuals.

Private information is gained by person-to-person contact—such as that between parent and child or possibly between the individual and the social group to which he belongs. The role of parents may well be crucial in that the encouragement or dissuasion they give to their children may affect the latters' achievement desires. It is thought, for example, that early 'mastery training' (teaching the child to do well, to act on his own, and to try hard) promotes high n-achievement *provided* the parent does not become too restrictive or authoritarian (Winterbottom, 1953). It does not necessarily follow, however, that parents who have themselves particularly strong achievement motives produce children of the same ilk; indeed, the reverse is the case: a very high level of n-achievement in parents is likely to lead to a lower n-achievement in their sons and daughters.

The sources of public information, in this context, include all the means of communication used by the individual. Folk tales, children's stories, comics, newspapers, films, radio and television shows—all may contain achievement imagery in varying degrees, some of which will certainly serve to condition the attitudes of their adherents (McClelland, 1961). Belief patterns form a part of this public information and, since they incorporate only such information or ideas as are deemed to be 'appropriate' or 'proper', they may cause the individual to regard all other information as irrelevant or, at best, as only marginally important. Thus, the Protestant ethic placed most emphasis on ascetic ideas which, in turn, became the major information field of its followers, in just the same way that in primitive societies taboos of various kinds result from emphases on certain ideas of the belief pattern. There is even some evidence to suggest that there are significant differences in n-achievement levels among different belief groups. McClelland, for instance, has shown that, other things being equal, Protestant boys of a given background usually have higher levels than Catholic boys of the same background (McClelland, Sturr, Knapp and Wendt, 1958).

Social class is a further determinant of motivation because it provides the individual or group with a particular set of values that effectively lead to the definition of their dominant information fields. In this respect, the role of social class is similar to that of belief systems. Indeed, the tendency (noted above) for high n-achievement to result from early 'mastery training' is very closely associated with social class origin. Lower class families tend either not to begin mastery training early in the lives of their children or, if they do, they tend to become too authoritarian and restrictive, with the result that low n-achievement scores are more frequently found in the children of lower class families. Similarly, social class may be even more significant than the 'native ability' or basic intelligence of individuals in determining the strength of their achievement desires. It has been frequently demonstrated in surveys of American high-school children that 'although intelligence is important, parental social status provides more motivation for high-school students to attend a university' (Lipset and Bendix, 1964, p. 230). Thus, the percentage of children of any I.Q. level who wish to attend further-education courses is markedly lower in the families of low social or occupational status than it is in families of higher status (White, 1952; table 2.8).

From such considerations of the information sources available to the individual, it follows that achievement motivation is not simply induced either genetically or environmentally. This is not to deny, however, that such factors may have an indirect effect. Indeed, Cortés (1960) has shown that high levels of n-achievement are positively associated with mesomorphic physique (muscular, thickset, large build) and negatively correlated with ectomorphic physique (delicate, light build); but this is usually taken to indicate an *effect* rather than a *cause* of achievement level. It can also be shown that achievement levels are highest in societies located in temperate latitudes; but again this is usually interpreted as an effect rather than as an ultimate cause.

There is little doubt that strong desires to achieve, however acquired, are

TABLE 2.8 ACHIEVEMENT AND SOCIAL CLASS: percentage of sample students wishing to attend further education courses, Exeter, 1972.

Social class of parents (Registrar General's classification)	I.Q.		
	<101	101–115	>115
1–2	78	80	81
3–4	68	68	69
5	47	49	58
8	31	37	45
9	28	32	45
10	15	26	39
11	8	11	24

responsible for success in the life of individuals or groups. Managerial success in general increases in proportion to the individual's n-achievement rating (though there are certain important qualifications to this rule). And the rate of economic development of society appears to be very closely related to the acquisition of strong achievement desires on the part of that society; periods of economic expansion usually follow periods of increase in achievement imagery in the literature and culture of society. It is this general conclusion that has led many observers to suggest that the foremost prerequisites for economic growth of underdeveloped countries are the quickening of achievement motivation generally and the rise to power of a class of entrepreneurs with similarly strong achievement desires (Lewis, 1955; Rostow, 1963; sections 5.2.2 and 6.1.2).

2.3.3 *Aesthetics*

To greater or lesser extents, most individuals appear to possess some sense of the aesthetic which, in part, affects their judgements and attitudes. Philosophers find it notoriously difficult to define the aesthetic, probably because the criteria by which it can be assessed are extremely varied, not only in themselves but also among different individuals (Charlton, 1970). Aesthetic judgements are little other than expressions based on the individual's intuitive feelings for and reactions to different situations. As such, it becomes evident that the structuring of preferences is inextricably associated with aesthetic values. The individual's choice of what is pleasing, attractive or repulsive in this aesthetic sense may well become more important than any other consideration in his choosing among alternatives. In certain circumstances, aesthetics may even condition the fundamental motivation of that individual, as Rooney discovered in analysing reactions to the snow hazard in the United States: 'the aesthetic values of snow make us somehow reluctant to wage all-out war against it' (Rooney, 1967, p. 556).

The term 'somehow' is indeed appropriate for, as yet, little is known of the precise way either in which these values take effect or in which they are determined. Even less is known about the important factor of 'spatial aesthetics', that is, the way in which individuals react to spatial situations, or indeed help to create them.

In just the same way that social values seem to result from a combination of attitudes scaled on different 'dichotomies', so aesthetic values appear to reflect the individual's view of a number of dichotomous variables relating to his perception of spatial arrangements. Beck (1957), for instance, has shown that, in expressing preference among different spatial designs, individuals may take into account five such dichotomies: the up–down, right–left, vertical–horizontal, delineated–open, and diffuse–dense aspects of space. Children apparently first appreciate only the left and right and the up and down qualities of spatial design (Cassirer, 1953; Piaget and Inhelder, 1956); but spatial differentiation tends to increase with the age of individuals and is associated with an increased awareness of the diffuse–dense aspect of space. In other words, experience affects the individual's aesthetic sense. It also appears that aesthetic values are in some way related to the achievement motivation level of individuals, though the exact form of the relationship is by no means clear. The researches of Aronson (1958) show that there is a tendency for schoolchildren with high n-achievement levels to prefer and create designs that tend to include 'discrete', S-shaped and diagonal lines with little blank space left at the foot of a page. Children of low n-achievement conversely appear to prefer 'fuzzy' diagrams with more scribbles, multiple curve lines and more blank space. In a similar way there appears to be some correlation (though rather tenuous) between achievement levels and preferred colours. High achievement desire is associated with preference for blues and greens rather than reds and yellows; and this could be interpreted as being the result of hard personalities requiring a 'soft' environment that they can dominate (Knapp, 1958).

The 'geometrical' qualities of the position and form of space thus underlie the individual's sense of the aesthetic, which is manifested in many different ways. Taste is one of the main characteristics stemming from aesthetic values, and to that extent it can be suggested that the architectural styles of different eras may well be conditioned by the prevailing 'taste' of the generation. It is true that the function of any particular building is usually considered when it is being designed, but it is also the case that 'our notion of what is "proper" in the shape of a building (that is to say, beyond the form which satisfies our need for shelter, warmth and the accommodation of the principal private and social acts) is governed by the powerful force of taste, a nebulous compound representing the cumulative influences of moral and aesthetic judgements on what makes a good appearance' (Johns, 1965, p. 4). In this way, it might appear that Victorian taste, for example, 'was confused by the belief that ornament and design were identical and was profoundly influenced by a secret religion . . . the religion of comfort' (Gloag, 1962, p. xv).

The taste and fashion of each age is naturally affected by all that has preceded it (once again, the familiar 'chicken and egg' argument emerges), so that even 'reactions' may well be aesthetically determined. Present-day townscapes and landscapes are consequently amalgams of styles and shapes representing the changing tastes and fashions of different ages. It can be argued that the Englishman's liking for the picturesque is nothing more than 'a preference for the irregular, the complex, the intricate, the ornate' (Lowenthal and Prince, 1965, p. 192), and this in turn is based on the geometrical arrangement of shapes into spatial arrangements in the 'up—down', 'left—right', 'vertical—horizontal', 'delineated—open' and 'diffuse—dense' planes. Even the arrangement of an individual's garden tends to be an expression of his appreciation of the symmetrical or asymmetrical. From a trial survey of garden designs, Johns (1969) was able to show that asymmetrical design was most common in the gardens of detached houses. It is an interesting speculation whether this may indicate 'that a connection exists between asymmetry and the individuality or creativity of the gardeners' (Johns, 1969, p. 56).

It is difficult to ascertain whether there is ever a true concensus of opinion in a given population regarding the aesthetic; yet, in general terms, it has often been suggested that the English tend to prefer the irregular and romantic to the geometric and formal in the landscape. Their alleged passion for the bucolic, the picturesque and the tidy, manifested so often in building camouflage (façadism) may well be a factor that should be borne in mind by present-day planners who, to a considerable extent, have so far managed to produce only what are aesthetically the opposite kinds of buildings and landscapes.

2.3.4 *Satisfaction*

The individual's view of what he wishes to achieve normally embodies his concept of what constitutes the 'most satisfactory' state of affairs. Naturally, this concept varies quite considerably among individuals, from those who always strive for the 'best' possible alternatives to those who will be satisfied with an alternative that may not necessarily be the very best possible.

Traditionally, economic theory (and a good deal of locational theory) has assumed that man behaves essentially as an 'optimiser', that is, that he will seek the course of action that will yield him the maximum possible benefit. Such benefits may be measured either in terms of money (maximum profit, minimum cost), risk (minimum risk) or satisfaction (maximum pleasure, least discomfort). Clearly, if such maxima or minima are to be achieved, the decision maker must have complete and totally accurate information about all the various alternative actions and outcomes. In other words, it is necessary to have a perfect evaluation of all alternative utilities. As has already been demonstrated (section 2.1), it is unlikely that an individual decision maker would ever have such complete information (because of time, cost, search or diffusional processes), so that optimised behaviour will inevitably be a rare phenomenon.

It may be that sub-optimal behaviour becomes typical of individuals because of their lack of complete information, in which case it may seem appropriate to consider them as 'born optimisers, saddled with ignorance or illfortune' (Kates, 1967, p. 63). On the other hand, it may be the case that, over time, decision makers become quite content with such *satisfactory* outcomes and eventually make all their decisions on the basis of satisfaction rather than optimisation. In either case, as Simon (1957) has pointed out, the result is the same: most decision making becomes concerned with the discovery and selection of *satisfactory* alternatives, and only in exceptional circumstances is it concerned with optimal alternatives. The exceptional cases normally concern large organisations rather than individuals. Having greater financial and other resources than the individual, governments, corporations and firms may be able to seek optimal solutions and consequently behave more as the 'optimisers' of traditional economic theory. Certainly, most firms now have sizeable operational research departments whose job it is to further that end. Once again, therefore, differential motivational characteristics may typify decision-making systems at different resolution levels. In reality, most individuals, while striving for the best possible outcome, rarely achieve it and usually have to settle for an outcome that is merely satisfactory and consequently less than optimal. The extent of this sub-optimal behaviour has been suggested by Wolpert (1964) in a study of farming in central Sweden that compares the actual productivity of labour with the optimum productivity that is theoretically possible. The difference between the actual and the theoretical is so great (even allowing for any gross exaggeration in the calculation of the theoretical figures) that it is obvious that in reality farming behaviour is widely sub-optimal (figure 2.10). Similarly, in studying man's response to natural hazards, Kates has argued that 'the level of adjustment is often sub-optimal: that is, fewer and weaker steps are taken than are required to minimise the effects of the natural hazard, while permitting the maximum use of resources associated with that hazard' (Kates, 1967, p. 61).

Underlying the search for satisfaction is the assumption that the decision

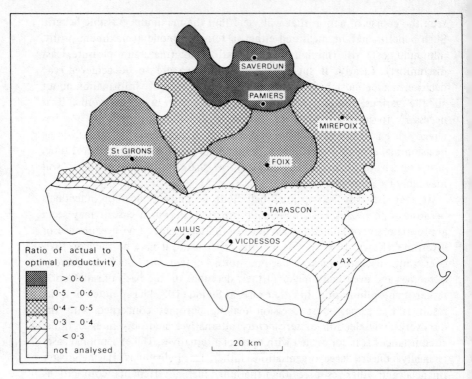

Figure 2.10 SUB-OPTIMAL LAND USE ratios of actual to optimal productivity based on the techniques outlined by Wolpert (1964). (Ariège, France.)

maker behaves rationally, that is to say that he can order all the outcomes and preferences, and choose on the basis of rationality which alternative yields the best or the most satisfactory result. Both 'optimiser' and 'satisficer' behaviour are essentially rational in this sense; but it is quite clear that in certain situations decisions may be taken that are based more on whim, fancy or chance than on rationality. Such irrationality appears to characterise many locational decisions. Most of us have probably remarked at some time or other that a particular firm or shop seems to be in the wrong place; and it is true that failure in business is frequently due to eccentric location of this kind. Furthermore, decision makers often behave *inconsistently,* and are liable to make different decisions at different times, even though all the conditions of information and motivation remain the same. In a very real sense human decision making is 'neither wholly rational nor wholly chaotic, but a probabilistic amalgam of choice, calculation and chance' (Haggett, 1965, p. 27).

The decision maker is consequently the cornerstone upon which the control system is founded, and as such is the most significant element of that

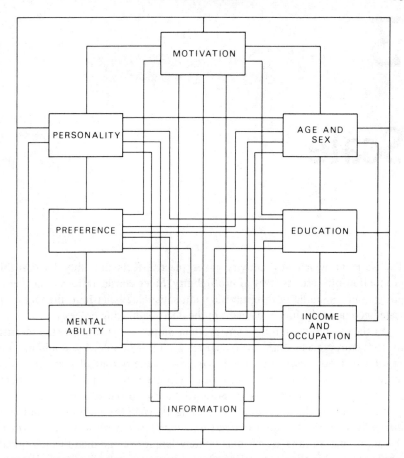

Figure 2.11 THE DECISION-MAKING SYSTEM direct, looped, series and parallel relationships between the personal characteristics, information, preferences and motivation of decision makers.

system. At the same time and at a different resolution level, this same element can be viewed as a complex subsystem of direct, indirect, series, parallel and feedback relationships among the personal characteristics (age, sex, education, income, occupation and mental ability), information, preferences and motivation of different individuals and groups (figure 2.11).

FURTHER READING

Brookfield (1969); Day (1969); Downs (1967); Edwards and Tversky (1967); Golledge and Brown (1967); Hall (1966); Hamilton (1971); Lichfield (1971); Lynch (1960); Parry-Lewis (1971); Pearce (1971); Pred (1967); Riggs (1962); Sopher (1967); Thompson (1967); Townroe (1969 and 1972); Tuan (1968); White (1967); Wood (1970).

3

Scale

The scale at which any activity takes place affects not only its cost of production but also its level of profitability. Since considerations of cost and profitability underlie both optimising and satisficing behaviour, the factor of scale becomes a significant element in the decision-making process.

As the scale of operation in any production process changes, so the returns to the entrepreneur may also change. If, for example, by doubling the amount of labour used for a particular process the output of that process is more than doubled, the returns to the entrepreneur in terms of extra product are clearly increased more than proportionately to the additional inputs of labour required. This situation of *increasing marginal* (extra) *returns* contrasts markedly with the situation of *decreasing marginal returns* which arises when the additional outputs are proportionately less in volume than the additional inputs incurred in their production. Such a situation may arise simply because too much of a particular input (such as labour) is being applied. Between these two extremes the condition of *constant marginal returns* arises when the additional output created is exactly equal to the additional inputs needed so that the raturns to the entrepreneur remain constant (figure 3.1A).

The cost of producing each unit of output (the 'unit cost') is closely related to the conditions of marginal returns that characterise the production process. In a situation of increasing marginal returns, unit costs are lowered because the total costs of production are spread over proportionately more units of output. Conversely, unit costs can only increase when decreasing marginal returns characterise the production process, but when production is continued under conditions of constant marginal returns, unit costs also remain constant (figure 3.1B).

The production processes of most firms and industries are characterised by all three of these alternative conditions of scale returns. At certain levels of

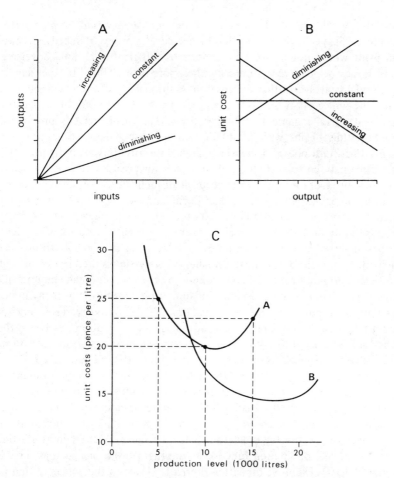

Figure 3.1 MARGINAL RETURNS A–output and input; B–unit costs and output; C–line A, the production function of a small firm; line B, the production function of a larger firm.

output it may be possible to derive increasing marginal returns, but as production increases, the stage of constant marginal returns may be reached and eventually decreasing marginal returns may set in. The levels at which scale conditions change vary from firm to firm and from industry to industry depending on their relative sizes, and consequently on their overall operating costs. Consider, for example, the production possibilities of two breweries each of different size (figure 3.1C). A small independent brewer might find that with existing plant his unit costs could be brought down to 20p per litre if he could produce 10 000 litres per day. A production level of anything less than this would inevitably mean that the unit cost would be higher (for example, 25p per litre if only 5000 litres per day were produced), because the plant would not be working at its most efficient level; but conversely, while it may be *possible* to produce rather more than 10 000 litres per day, to do so might involve further expense on additional inputs such as labour, and cause the unit costs to rise (for example, at a level of 15 000 litres per day, unit costs may increase to 23p per litre). The small brewer's production function thus begins with a situation of increasing marginal returns, but finally turns into one of decreasing marginal returns (curve A in figure 3.1C). The same may be true of all other breweries for precisely the same reasons, but their production possibilities occur at different levels of output and unit costs because of their different equipment and overhead costs. A medium-sized brewery has more and larger vats than the small brewery so that it can produce more beer than the small brewery, but even if it operates at a level that is slightly below its optimum capacity it might still be able to produce each litre more cheaply than the smaller brewery simply because of its size. However, the point may well come where to produce only small amounts in a large plant would create a situation of diminishing returns (too many inputs per unit of output), with the result that *at low levels of output* a smaller firm might still have a comparative cost advantage of production relative to the larger one. The larger firm, however, soon achieves increasing marginal returns and is able to undersell its smaller competitors (curve B in figure 3.1C).

There are two ways in which the individual firm may be able to organise production in order to achieve increasing scale: either it may endeavour to increase its output with existing plant and techniques until the point of lowest unit cost is reached, or it may change its plant, organisation, techniques or location in order to achieve the same effect (thereby shifting from its original production function to another, lower, one such as curve B in figure 3.1C). Patently, the latter alternative leads to the concentration of production in a few large plants or firms. The result may be that industries become increasingly 'localised' and the distribution of plants much more 'sporadic', but the extent to which different firms and industries are able to make such changes varies considerably, as Bain (1954) has shown in a study of twenty American industries. In some, such as the cigarette manufacturing and rubber industries, scale appears to be a relatively insignificant consideration, whereas in others, such as the petrochemical industry (Isard and Schooler, 1955), the factor of plant size completely overshadows any other locational or cost consideration in the decision-making process (table 3.1). Industries or firms which require a great deal of personal contact with their customers are least able to benefit in this way since they cannot concentrate their production into one or two localised large plants without loss of contact. Nevertheless, as population mobility increases (section 5.4), this

TABLE 3.1 UNIT COSTS OF PRODUCTION AND SCALE OF PLANT: selected American industries, 1947. Source: Bain (1954). Base of 100 applies to costs in the largest size plants (that is, those whose output exceeds 5 per cent of the total national output of all plants in the particular industry). Note how unit costs increase, generally, as size decreases.

Industry	Size (measured as % of national industry capacity contained in one plant)				
	5%	2½%	1%	½%	¼%
Cement	100	100	100	115	130
Distilled liquor	100	100	100.5	101	102
Petroleum refining	100	100	102	104	107
Tyres and tubes	100	100.3	103	104	105.5
Rayon	100	107	125	?	?
Soap	100	103	105	>105	—
Cigarettes	100	101	102	>102	—

obstacle to increasing size may be lessened or removed altogether. Certainly there is ample evidence to suggest that economies are now being made in many service industries: the emergence of supermarkets and, more recently, of hypermarkets and cash-and-carry discount warehouses, coupled with the consequent decline in small retail outlets represents the increasing derivation of scale economies in retailing (it is estimated that in America the number of retail stores is decreasing at a rate of 5.8 per cent per annum due to this factor alone).

Limitations to the possibilities of the physical expansion of plants is frequently imposed by the lack of available suitable land for expansion. If existing plants cannot expand physically on their present sites, relocation will be the only alternative possibility. But relocation may also be difficult because the number of sufficiently large sites that may be available in any area is obviously more limited than the number of small sites. Scale considerations may consequently cause an effective indirect restriction on the range of alternative locations available for industrial and commercial premises, and they may additionally lead to local and regional problems of unemployment and site dereliction.

The possibility of increasing output is limited not only by the size of plant and the availability of adequate machinery, land or labour input, but also by the effective demand that it is possible to supply at any location. In this respect, firms that are located in agglomerated areas of population should be better able to derive internal scale economies than those that are located in areas of sparse population, since the availability of a large population may necessitate greater output in order to meet the level of demand that it generates. It is largely for this reason that the unit cost of provision of such amenities as education, refuse disposal and other services provided by local authorities decreases significantly as the size of the urban area for which they are provided increases (Lomax, 1943; Duncan, 1956; Hepworth, 1970). Similarly, many firms in the distributive trades have discovered that unit costs can be reduced and marginal returns consequently increased, by large scale operation, and that one of the ways in which this can be achieved is by optimising delivery and collection routes within increasingly large trading areas (Cigno, 1971). Naturally, there are limits to the economies that can be derived from serving greater populations in larger areas as there are with all other sources of saving and, eventually, decreasing marginal returns begin to set in at various size levels for different services (figure 3.2; Isard, 1960; Gupta and Hutton, 1968).

The agglomeration of several firms and several industries at one geographical location represents a further, and very significant, form of scale economy. If competitors and dependent industries locate in close proximity to each other, the effect in the long run is essentially the same as increasing the market area or of reducing transfer costs, so that greater marginal returns can be derived. Again, however, these 'external' economies (so called because

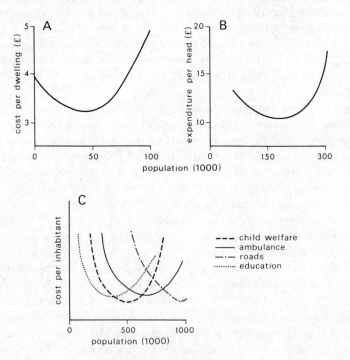

Figure 3.2 SERVICES AND SCALE (Britain, 1969) A—unit costs of housing; B—rate fund expenditure per head; C—cost of social and welfare services.

they are derived by processes external to the firm) may turn into diseconomies at certain critical size levels and may lead paradoxically to the decentralisation of economic activity.

By such mechanisms, returns to scale may be derived both internally and externally to individual plants and to industries as a whole and may necessitate a continuing re-appraisal of geographical location.

3.1 INTERNAL ECONOMIES

The extent to which scale economies may be derived within a particular plant is clearly dependent on the extent to which efficiency can be increased or maintained. A number of ways may be possible, but they are all based on the intensification of the use of plant and overheads to the point at which marginal returns begin to diminish.

3.1.1 *Division of labour*

The greatest possibility lies in the introduction of specialisation. The production process can be broken up into several separate parts either broadly, as between production, packaging, marketing and distribution, or minutely, as between all the processes involved in each of these broader divisions. In a small firm or cottage industry one man may be expected to perform all the different production operations right through from beginning to end, with the result that he becomes something of a 'jack-of-all-trades'. In a very real sense, however, he also becomes 'master-of-none' and probably wastes a considerable amount of time in passing from one job of work to another. By subdividing the process and letting different people specialise in different jobs, greater productivity may result because the dexterity of each person is increased and time that was lost in transferring from one job to another is saved. A man or woman who works continuously at one job for some considerable time is able to establish a rhythm of work that may lead to an almost automatic reflex action in doing that particular job, and that may enable the worker to maintain a high rate of output for a considerably longer period than would have been feasible without such specialisation. Additionally, specialisation of labour enables the best possible use to be made of available skills, aptitudes and talents. Management can be affected in just the same way as other operations: in a small works, the manager may have several different functions to perform, ranging from accounting to personnel management, whereas in a larger firm, the whole process of management could be subdivided into such specialist departments as finance, sales, buying, marketing or maintenance, each of which becomes more efficient in its own sphere.

The effect of such division of labour is that the large firm is able to secure an efficiency of work normally beyond the reach of the small firm, and the additional cost that may be incurred in setting up such an elaborate division of skills is more than offset by the resultant increased productivity. There are, however, limits to the benefits that may be derived, since it is possible to become over-specialised. Every time a further division of labour is introduced, the problem of co-ordinating the work of the separate groups begins to arise. Colloquially it might be said that the right hand does not know what the left is doing—and most of us have heard that criticism laid against large organisations such as universities, the Civil Service, local government, the Church and many industrial and commercial firms. Indeed, it has been repeatedly argued that the point of diminishing marginal returns will be more rapidly reached through over-specialisation and division of processes, and just how large a firm can successfully grow 'will depend upon how it solves this problem of the co-ordinating of separated departments and separated specialists' (Robinson, 1958, p. 25).

3.1.2 *Substitution*

Further economies can often be gained by substituting machines to do the work of men or women: that is to say, capital is substituted for labour, but the extent to which any economy may be derived from this form of factor substitution is dependent on the costs of the two inputs concerned with respect to the marginal product which they create. Only if relatively more product is created through the substitution of one unit of capital cost for an equivalent unit of labour cost, does the substitution produce any economy.

Obviously, a large firm is better able to substitute expensive machinery for labour than is a small firm, since it can afford to buy machinery which, for a smaller scale of operation, would be so costly as to be unprofitable. Furthermore, it is only the large firm, with its larger quantity of output, that can keep the expensive machinery working near to its maximum capacity. To work machines at less than capacity can only increase the unit costs of production and lead to a situation of diminishing marginal returns.

Large firms are also better able to borrow the necessary capital for this kind of substitution. A larger firm's name is already better known to the investing public and its standing can be more easily ascertained, with the result that it can attract capital for development more easily than a small firm (see also section 6.2.1).

3.1.3 *Disintegration*

The small firm can, however, obtain many of the advantages of specialisation and substitution by disintegration of their processes. A number of small firms can subcontract certain processes to an outside specialist firm that, because it supplies all the separate small firms, is working at a larger scale than any of the individual firms could have achieved for that particular process. The newly-disintegrated firm is then able to achieve increasing scale economies by virtue of its size. This same process may equally apply to certain parts of the processes of large firms, and it is for this reason that disintegration is evident in many industries, especially in textiles and in the motor industry where several components—such as car bodies, instruments or brake linings—are made by subcontracted firms. In agriculture, specialist contracting firms have arisen in the spheres of plant hire and artificial insemination to take advantage of similar economies. In retailing, disintegration occurs through the banding together of small shopkeepers into larger trading groups such as Mace, Vivo, Spar or Centra with the central organisation placing bulk orders, not only for branded supplies but also for their 'own-label' products.

Disintegration may also lead, somewhat paradoxically, to the financial merging or takeover of the disintegrated firms because the dependent firm may feel that it should retain *some* control over its supplies. The Pressed Steel–Fisher Company which supplies car bodies to British Leyland is now, for instance, under the financial control of the latter company (section 9.1.2).

3.1.4 *Balance of processes*

Every part of the production process has its own optimum level of operation at which unit costs are minimised through specialisation, substitution or disintegration. Once it has recognised the optimum levels characterising the different parts of its production process, the management of the individual firm is faced with the difficult task of deciding how to equate production levels of all the processes. Very often the optimum levels are incompatible with each other (figure 3.3): a firm may well find that to adopt its optimum size in terms of management would entail the adoption of a technical and marketing output that would be less than optimal. In such cases the firm must either seek a compromise solution based on some kind of Highest Common Factor of each process, or else it may seek a merger with other firms, or further disintegrate parts of its processes. It is for this reason that mergers and disintegration are the inevitable results of the search for internal scale economies within different industries. The geographical consequences of mergers and disintegration may be profound in that the locational requirements of the newly-structured firm may be very different from the conditions that had led to the location of each of a number of formerly separate firms acting independently of each other. Locational revaluation, with all its implications, may thus be the ultimate result of the derivation of internal scale economies.

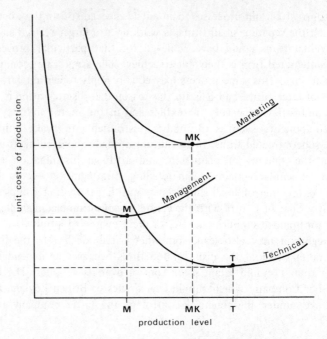

Figure 3.3 INCOMPATIBLE PRODUCTION SIZES hypothetical firm in which maximum technical scale economies may be achieved at a production level T, maximum management economies at level M and maximum marketing economies at level MK.

3.2 External Economies

Just as a single firm can achieve internal economies as it expands its output and increases its efficiency, so a whole industry or several different industries may be able to achieve external economies by means of agglomerating together spatially. The economies to be derived are not only related directly to the physical size of the agglomeration but also indirectly and fundamentally to other advantages of convenience resulting from competition between firms in a spatial setting.

3.2.1 *Competition*

It may seem paradoxical that competing and rival firms are not infrequently located in close geographical proximity to each other, but so great are the economies that can be derived from spatial agglomeration that other considerations may be dwarfed into insignificance.

Consider a large beach, uniformly peopled with sunbathers on a not too sunny day, and suppose that a single ice-cream vendor comes along to sell his wares. In such a monopolistic situation the vendor could locate anywhere he liked since, if people wanted ice-cream, they would have no alternative than to get it from him wherever he was located. On the other hand, it could be argued that it might be in the vendor's best interests to find a site that was central to the whole beach so that all his potential market was within easy reach of him. In such circumstances, people from the far extremities of the beach might be tempted to go for a relatively short walk to the centre, whereas if the ice-cream vendor were located at the opposite end of the beach they may decide that the walk was just not worth the effort and remain untempted all day. The monopolistic vendor thus decides to maximise his sales and locates himself in the centre of the beach (point X in figure 3.4A).

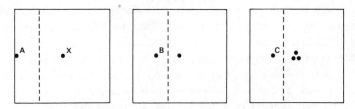

Figure 3.4 SPATIAL COMPETITION the ice-cream vendor problem.

A second ice-cream seller now arrives on the scene and, it may be assumed, is concerned to maximise his takings for the day. Where will he locate, now that the plum central site is already occupied? It might be argued that he should locate some distance away from his competitor so that he could establish a sort of 'mini-monopoly' in his own area. If he were to set up stall at the far end of the beach (point A in figure 3.4A) he should be able to take a fair proportion of trade away from his rival; however, sunbathers will normally tend to go to the nearest vendor, so that he can only hope to attract those sunbathers from distances up to halfway between his rival and himself (shown by the dotted line in figure 3.4A); in other words, he can only hope to gain a 25 per cent share of the total market by locating at that position. Clearly, he may be able to do better than this by moving nearer to his competitor: suppose that he moves to a point half-way between the edge and the centre of the beach (point B in figure 3.4B): at this point his share of the trade rises to 37.5 per cent (that is, three-eighths of the area). By similar reasoning, it becomes obvious that if he were to locate next door to his rival *right at the centre of the area,* he should certainly be able to increase his share of the market to at least 50 per cent. The actual share of the market will be determined by such characteristics as the reputation of his product compared with that of his rival but, in theory at least, if the products are broadly comparable, there is an obvious advantage to be gained from locating in close proximity to the competitor. The scale of operation of ice-cream selling at point X is now increased (by 50 per cent) and it is for this reason that the economy of proximal location is effectively a form of scale economy.

In reality, the successive agglomeration of competitors may lead to other advantages than those of mere ability to increase their share of the market. The area may begin to establish a reputation for its range of alternative sellers and producers; it may develop a certain expertise; and it may also begin to attract a range of specialist ancillary services that improve trading conditions generally (chapter 3.2.2). In this way, not only are better trading possibilities created for firms and industries, but more convenient conditions are created for customers who have the advantage of being able to 'shop around' at the one central location. Of course, the derivation of these scale advantages is fundamentally related to the continued accessibility of the central location with respect to its market area. Indeed, it is the relative accessibility of alternative locations that makes them potentially desirable in the first place—the ice-cream seller's assessment of the centre point of the beach as the 'best' location is based on the assumption that that point is the most accessible to all customers. Different industries and enterprises naturally define 'accessibility' in rather different terms and a site that is considered to be 'most accessible' for one activity may not necessarily be the most accessible for another. However defined, it is accessibility that determines the initial location of the first firm in an area through its ability, relative to other competitors, to bid for what it considers to be the most accessible site (section 7.3.2). Thereafter, agglomerative economies combine with accessibility to produce a centripetal tendency for competitors to locate together and to produce distinctive spatial zoning or sectoring of economic activity.

Industrial sectors in urban areas thus reflect the continued spatial agglomeration of competing firms at locations that once possessed some initial advantage of 'relative accessibility'. The jewellery and gun 'quarters' of Birmingham, the clothing and furniture 'quarters' of east London and the cutlery 'quarter' in Sheffield are all examples of the general effect, but most urban areas are characterised by the sectoring of industrial activity at characteristic locations (Lowenstein, 1963; figure 3.5).

Agricultural 'belts' may also be as much the result of the derivation of scale economies as of the physical conditions that may be necessary for production of any given kind. The way in which Kent became the specialist area of hop cultivation in Britain can, for instance, be fully explained in terms of the external economies of scale that accrued from establishing new acreage in the existing locations that had gained a reputation in both cultivation and marketing (Harvey, 1963). Similarly, there is ample evidence (figure 3.6) of the functional grouping of similar retail activities in any central business district of a town or city. Food shops, for instance, are characteristically

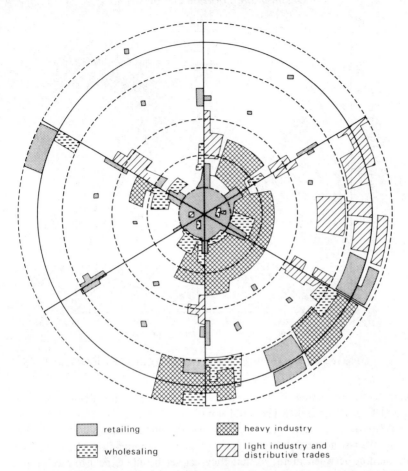

Figure 3.5 LOCATION OF INDUSTRIAL ACTIVITY a general model of the sectoring of industrial activity within urban areas, based on empirical data of 24 British cities.

located together near the edge of the CBD, while clothing stores tend to cluster together at more central locations (Murphy, Vance and Epstein, 1955; Getis, 1968).

As with all scale returns, there eventually comes a point at which the original economies (advantages) become less marked and diminishing returns may set in. This situation can be simply demonstrated once more in terms of the ice-cream vendors. Suppose that three vendors had established themselves at the centre of the beach and that they had each been able to take 33 per cent of the total market. A fourth vendor might now find that he could do better by locating eccentrically away from the established centre. Suppose he

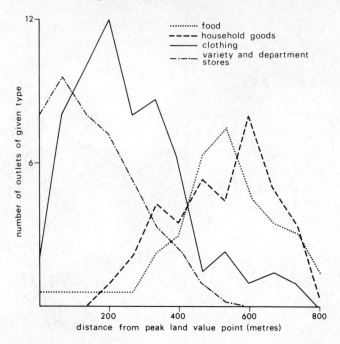

Figure 3.6 RETAIL OUTLETS IN THE CBD City of Exeter (1972).

were to set up stall halfway between the centre and the edge of the beach (point C in figure 3.4C). He could now hope to attract 37.5 per cent of the total market at that location (for reasons already analysed). His three competitors find that they now have to share the remaining 62.5 per cent of the market between them, so that they can each only expect to receive 20.8 per cent of the total market. Had all four been at the centre, they could each have hoped for 25 per cent of the market. It may be that sunbathers will prefer to go to the original centre with its greater potential *choice*, but it is clear that this is the crucial stage at which economies of agglomeration are beginning to be outweighed by the possibilities of deglomeration.

In addition to the problem of the possible decreasing size of the share of the market, account must also be made of the increasing congestion in the central area as more and more competitors move in (and as other activities also move in—see below). In reality, an increased number of firms trying to seek out similar proximal sites tends to make the cost of the sites rise since similar sites are in very limited supply (section 4.2.2). Equally, other costs may increase in the central area due to increased competition (wage rates may be forced up for instance (section 5.3). When this happens, diminishing marginal returns occur since unit costs of production are also increased; the net effect of which is to cause new competitors to locate elsewhere than at the centre and also—if the diseconomies become very great—existing firms may attempt to relocate away from the centre. Diminishing marginal returns thus have a centrifugal or deglomerative spatial effect—a tendency that is manifest in the migration of industrial firms from central sites to peripheral trading estates, the development of out-of-town shopping centres away from the CBD, and the extension of agricultural belts. Harvey, for instance, has shown how such diseconomies began to typify the central areas of the Kentish hop industry when the rent of land and the cost of hop poles became so high that unit costs began to increase. The result was that new areas of production tended to be located elsewhere, but, 'the agglomerative pull meant that this relocation always took place with reference to distance from the centre. The interaction between the processes governing concentration and the centrifugal effects of diminishing returns can thus be regarded as the key explanation of the zoned distribution of hop acreage' (Harvey, 1963, p. 137).

3.2.2 *Cumulative causation*

Such centripetal and centrifugal forces resulting respectively from spatial economies and diseconomies of scale apply not only to competing firms within any given industry, but also to any form of economic activity. The reason for this is that as a location begins to develop as a centre for a particular activity, other activities will find that they can benefit from being in close proximity to those original industries. Consider once more the continuing agglomeration of ice-cream vendors on a beach. It would clearly be to the advantage of sellers of any other product (such as peanuts, sunhats, suntan lotion or buckets and spades) to seek a central position on the beach for precisely the same reasons as it was to the advantage of the original ice-cream vendors. In other words, once a particular activity is initially established at a particular location there is every good reason why other activities should follow suit and locate there as well. The net result is that agglomeration of all activities will occur at locations that, for various reasons, seem to possess some form of initial locational advantage.

This initial advantage may, as in the ice-cream vendor problem, be related to the locational accessibility of a particular site, but it may equally be due to some simple historical accident (such as Lord Nuffield developing his car industry in his home town), or some other 'advantage' related to the physical characteristics of site (for example, a spring line or bridging point). Whatever the initial advantage, agglomeration will inevitably continue, though its extent may depend on the kinds of activity that are attracted to the location.

Among the activities that will follow initial industrial development are a host of ancillary industries that arise to serve the needs of both industry and population alike. Banking, insurance, repair and maintenance facilities all move in because they wish to secure as large a part of the market as possible. Furthermore, the individual plants of the original industries may also have disintegrated parts of their processes so as to derive internal economies. The entrepreneurs of the disintegrated plants find that their locational advantage is to be found in close proximity to their buyers who are, of course, mostly to be found in a limited quarter or spatial zone.

As economic activity continues to agglomerate, it becomes relatively cheaper to provide the basic physical infrastructure and amenities of the area. As has already been shown, the provision of education, water, sanitation, power and transport can all benefit from internal scale returns, so that the agglomeration of industry and population itself makes possible the cheaper provision of basic services. In this way, agglomeration begets agglomeration and urbanisation proceeds as the inevitable result of the operation of cumulative internal and external scale economies (figure 3.7). 'Urbanisation is not inexorably associated with industrial growth: yet, conversely, the multiplication of factories, product output and markets since 1860 is virtually synonymous with city development' (Pred, 1965, p. 158). As the agglomerating area continues its upwards spiral of growth it begins to draw to itself labour, capital and other commodities from surrounding areas (sections 5.4.1 and 6.2.1). People discover that opportunities appear to be greater in

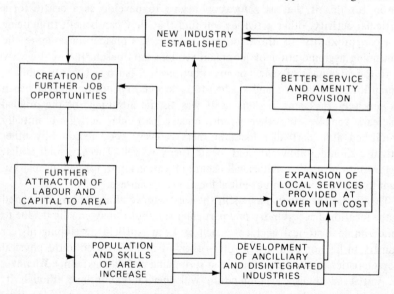

Figure 3.7 CUMULATIVE CAUSATION (POSITIVE) based on Myrdal (1957) and Pred (1965).

these expanding urban areas and, whilst perhaps realising that the streets there are not necessarily paved with gold, may be tempted to migrate away from their local area in search of greater prosperity. When this begins to happen, the process of cumulative causation begins to work in a negative form in the surrounding areas. As the population moves away, local services or industries can no longer be supported, so they close down and this in turn only helps to encourage yet more people to move to the imagined utopia of the developing agglomeration (figure 3.8). Agglomeration therefore 'polarises'

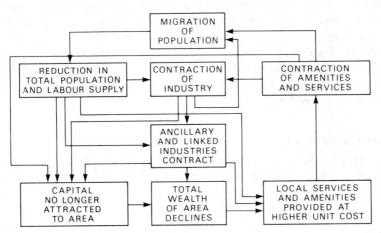

Figure 3.8 CUMULATIVE CAUSATION (NEGATIVE) based on Myrdal (1957) and Pred (1965).

development into a few localities and causes 'backwash effects' in the surrounding areas (Hughes, 1961; Lasuen, 1962) so that the processes of cumulative causation appear to 'call forth countervailing changes which move the system in the same direction as the first change, but much further' (Myrdal, 1957, p. 13). The ultimate effect of scale economies is thus to induce positive feedback between the elements of the system (section 1.1.2).

Eventually, however, a stage in the process of industrial agglomeration is reached where maximum economies of scale are achieved, and further expansion beyond that point only results in the inevitable diseconomies of diminishing marginal returns. Competition between firms and between industries eventually forces up the prices of materials and the factors of production (land, labour and capital) so that unit costs begin to increase. Similarly, diseconomies in the provision of rate fund services and other overheads occur, the total effect of which is to cause decentralisation and relocation of activity. The moving of various activities and population from central London to new growth areas (mainly near Basingstoke and High

Wycombe) is just one of the many examples of this tendency. In many ways, this process appears to bring about some form of 'equalisation' in the agglomerative process, and it may be that the eventual diseconomies in the process of cumulative causation may induce a self-regulating mechanism of negative feedback into the locational system. The earlier tendencies that appeared to be of a positive feedback nature may simply be an erroneous impression of a process that is in fact of a converse nature. Opinion on this matter is, however, divided: those who believe the relationship to be one of positive feedback would argue that few firms would be willing to move out from an agglomerated area *voluntarily* and that the only way in which decentralisation takes place is by legislation or via long campaigns of indoctrination aided by substantial financial inducement from various government or other bodies (section 9.1.3). Such artificial interference is made necessary by the long run tendency for scale economies to give rise to positive feedback in the system.

FURTHER READING

Baer (1964); Borts (1960); Borts and Stein (1964); Boudeville (1966); Chisholm (1963); Eckaus (1961); Friedman and Alonso (1964); Pred (1966); Williamson (1965).

4
Land

The relationship between man and the land is perhaps the most critical aspect of the spatial organisation of the system since it necessitates the resolution of an ever present dilemma created by the countervailing effects of scarcity and growth.

Scarcity is a diagnostic property that arises because land, unlike all the other factors of production, is in finite supply but is needed by all forms of human activity both for the physical sites that it provides and for the many resources that it contains. Yet growth and development are diagnostic properties of all systems, and increasing population and economic prosperity merely lead to increasing and often conflicting demands for land.

Land which is needed for one activity may be equally needed by other activities. Housing development, for instance, normally requires well-drained and flat (or, at most, moderately sloping) land in close proximity to existing or future places of employment and service provision. In other words, sites that are on good land and peripheral to urban areas are most in demand for residential purposes. Yet it is precisely that kind of land that is needed for agricultural, industrial, commercial, transport and recreational purposes. Similarly, conflicting demands arise from preservationist attitudes: cries of 'hands off' some particular piece of amenity land, because it is needed for development, are often heard as the land-use conflict increases, but to call 'hands off' in one place entails the calling of 'hands *on*' somewhere else. The proposal to construct Swincombe reservoir on Dartmoor was opposed in this way by several interested parties, but the result of their successful pleas was to transfer the problem to the alternative site in the agriculturally prosperous South Hams region of South Devon—a proposal that was just as strongly opposed by the residents of that area. The land use conflict is hardly likely to be effectively resolved if such sectionalist and preservationist attitudes are allowed to prevail at a time when it is apparent that the imbalance between demand and supply of land may become acute within a very short time in many parts of the world. Cracknell (1967) has shown that by the year 2000, the situation in England could well be serious. Out of the total land area of 129 645 square kilometres, 41 906 square kilometres are already used for residential, industrial and recreational purposes, so that only 87 739 square kilometres remain available for future development. However, if the land that is considered to be unsuited to residential development, and that which is of great agricultural or amenity value, is subtracted from this total, only 65 946 square kilometres remain (figure 4.1). If it is assumed that population growth will create an additional 23.7 million people by the year 2000 and if it is further assumed that for every 1000 population, 3.24 square kilometres of 'living space' are required, some 77 182 square kilometres will be needed for residential and living purposes. In effect there would be a shortage of 11 236 square kilometres of land. Although such alarming projections may not typify

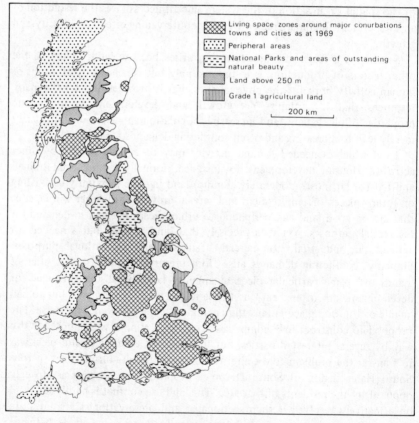

Figure 4.1 LAND OCCUPANCE Britain. Source: Cracknell (1967).

all systems at the present time, there can be little doubt that *potentially* similar situations may result if demand outstrips the supply of the one finite resource of the system. It is therefore essential that land should be properly allocated between alternative uses and developed without unnecessary waste and that growth should itself be effectively stabilised.

The need for the application of optimal decision making in this respect should be apparent, yet it has often been argued in the past that man's role as a decision maker can be almost completely disregarded because the use of land is determined by ecological conditions alone. Such a deterministic proposition is clearly at variance with reality for it denies man the possibility of any freedom of choice within his environment. It is true that ecological conditions are important determinates of land use, but there is an intricate and delicate relationship between ecological systems and man-made systems that tends to suggest that man is a decision maker 'who is very sensitive to variations in the land resource base, and who displays a strong rationality by reacting sensibly to them' (Found, 1972, p. 22). Sensible reactions do not, however, necessarily lead to optimal patterns of land use since they tend to represent decisions that are based on 'satisficer' rather than 'optimiser' principles, and it is probable that many decisions concerning the management of land have, in the past, been of this kind (figure 2.10; section 2.3.4). A rigorous analysis of the causes of the various demands for land, and an equally close evaluation of both the ecological and economic characteristics of land at different locations, is necessary if such satisficer behaviour is to be replaced by better land use decision making.

4.1 DEMAND

The amount and type of land that is needed at any time and at any location is basically determined by the prevailing level and pace of economic growth and development; the extent to which existing structures are beginning to decay or become obsolete; and by the possibilities of intensifying both existing and future land uses generally. All three of these conditions are closely related to the level of technological skills that may be available and to the prevalent attitudes and motivation of decision making entrepreneurs in all forms of economic activity.

4.1.1 *Growth*

One of the consequences of economic development, and indeed the main purpose of it, is that standards of living are increased both in terms of wealth and health. Desirable as this may appear to be, it is also the fundamental cause of the land-use dilemma, since increased standards of living lead directly to greater demands for resources and for land on which to manufacture or produce the increased range of products required. Furthermore, changes in health and prosperity, coupled with the concomitant quickening of the 'desire to achieve' (section 2.3.2) lead indirectly to changes in population growth and structure that, in turn, create increasing demands on the land. In a very real sense, therefore, it can be argued that the only effective way of reducing pressure on the land is to 'stabilise' economic growth and development by making a complete re-assessment of the values and goals of society as a whole (Harvey, 1973). Whether this is a practical possibility, however, remains to be seen.

Growth of national income is related to the rate of capital accumulation and investment in the economy and while there are many initial difficulties involved in raising sufficient investment to generate a substantial growth of national product, once the 'take-off' point has been passed, economic growth appears to be more or less assured (section 6.1). Income levels consequently vary according to the general level of investment and pace of economic development in different regions at different resolution levels of the system (figure 4.2). As national income increases, so do the personal incomes of the individual members of the community: in the USA the average rate of growth of national income in the period 1945–70 was in the order of 1.9 per cent per

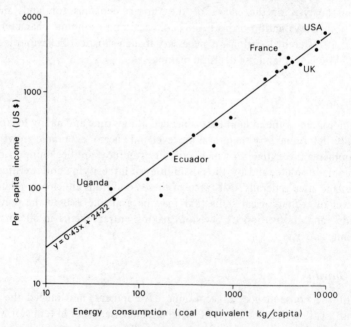

Figure 4.2 INCOME AND ECONOMIC DEVELOPMENT correlation between *per capita* income and economic development (measured by energy consumption).

annum, during which time *personal* incomes rose from $1355 to $2070 (an average rate of 2.1 per cent per annum); in Britain mean net personal incomes before tax have risen from £400 in 1949 to £1276 in 1972; the average rate of increase of disposable incomes in less developed countries since 1950 has, in many ways, been rather more significant since it is in those countries that *absolute* incomes have also been raised from previously very low levels (table 4.1). It is difficult to assess the extent of 'personal' income growth in the

TABLE 4.1 DISPOSABLE INCOMES, 1950, 1960 AND 1970. Source: *United Nations Yearbook of National Accounts Statistics* (1971), table 9.

Country	Currency	Per Capita Disposable Income		
		1950	1960	1970
Burma	Kyats	165.41	279.55	391.69
Colombia	Pesos	385.07	1 629.65	3 969.28
Costa Rica	Colones	1 259.98	2 291.27	2 597.11
Ecuador	Sucres	1 742.12	2 755.83	4 480.39
France	Francs	1 975.12	5 830.31	12 483.97
Malaysia	Malaysian $	410.33	793.03	978.92
Peru	Soles	1 605.62	5 764.21	12 738.80
Philippines	Pesos	275.07	481.65	662.51
Spain	Pesetas	9 900.63	17 331.85	46 579.86
Sweden	Kroner	3 509.27	7 911.87	15 002.53
UK	£	213.50	412.04	718.74
USA	$	1 532.62	2 419.71	4 502.45

future but it seems that at a world scale, the 1972 income levels will be doubled by the year 2030. Some nations will reach that level before then, others afterwards. In Britain, at least one estimate suggests that real income per head will have doubled in the period 1960–85 (Burton, 1967).

As incomes rise, demand for agricultural, industrial, commercial and other products also rises. Food consumption is often the first to increase consequent upon a rise in income: at the world level, the daily *per capita* intake of calories had increased from about 2100 calories in 1939 to just over 2500 calories in 1970, thereby achieving a standard that is commonly regarded as the minimum requirement for an average adult. By the year 2000, the total world consumption of food is expected to be in the region of 20 600–24 000 thousand million calories per day, compared with 7200 thousand million in 1960. Consumption of raw materials also increases as economic growth continues: technological innovation tends to create greater efficiency in the use of such materials, but at the same time more and more materials are needed for a wider range of purposes than previously, so that total demand still continues to increase. The consumption of raw materials for energy production, for example, has increased from 867 to 1635 kilograms of coal equivalent *per capita* per year in the period 1929–70 despite an estimated 100 per cent increase in the efficiency of their use. Estimates vary widely, but it seems probable that world energy consumption will increase from a 1960 world level of about 4.2 thousand million metric tonnes to a minimum level of 17.7 and possibly to a maximum level of 55.3 thousand million metric tonnes by the year 2000 depending on the rate at which income levels increase throughout the world (Fisher and Potter, 1964, pp. 40–7).

Demand for leisure activities also increases significantly as personal incomes continue to rise, since individuals are then able to spend more not only on food and other necessities but also on pastimes and recreation generally. Being able to spend more on such items, individuals attempt to 'substitute' leisure for work, and indeed two of the major concomitants of economic growth are that the length of the average working week is considerably reduced and the length of annual paid holiday is increased. Beyond some minimum critical level, expenditure and time spent on leisure tends to increase, often more rapidly than successive incremental increases in income so that the stage may be reached where the income elasticity of demand for pleasure is more than one' (Mansfield, 1969, p. 152; figure 4.3). Many leisure activities do not directly cause any demand for land since they are pursued in individuals' own homes or in the homes of friends and

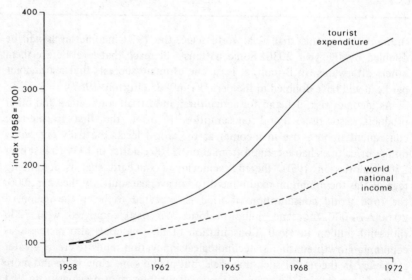

Figure 4.3 LEISURE DEMAND the growth of world national income and tourist expenditure 1958–72 (1958 = base of 100).

relatives. Watching television, listening to the radio, reading, decorating, maintaining vehicles and following up crafts and hobbies are all of this kind, yet they normally account for the largest proportion of leisure time—in Britain, 40 per cent of mens' leisure time and 50 per cent of womens' leisure was spent in this way (Sillitoe, 1969). Others, however, do place further demands on the land whether for the provision of gardens, buildings, play areas or holiday and sightseeing areas generally (Patmore, 1973; Toyne, 1974).

Gardening is an activity which may be very time-consuming as well as back-breaking; the average British male is estimated to spend 12 per cent of his free time in tending his garden and while the British may generally be regarded as rather curious if not actually quaint in their passion for gardening, there can be little doubt that gardening remains a significant leisure activity elsewhere. One of the most frequent criticisms of residential development is that the garden is either too small or is non-existent: Sillitoe (1969) found that 17 per cent of all males in Britain who already have a garden would like to have a larger one and there is every reason to suppose that a sizeable proportion of flat dwellers would probably also welcome the provision of this amenity. Yet the provision of private gardens is a very land-consuming process and it is only through the abolition of this facility that high-rise residential development is able to increase the density of land occupance (section 4.1.3).

Some leisure activities require land for the provision of buildings in which they can be pursued. Public houses, cafés, restaurants, clubs, libraries, hostelries, cinemas, bingo halls, theatres, dance halls, sports halls and other such facilities are regularly used by many sections of the community and have to be provided at least in relatively close proximity to residential areas. This in turn further increases the pressure on land that is available for residential development, yet public opinion tends to be such in developed societies that there appears to be little possibility of reducing the provision of these facilities in any general way. Rather, there is ample evidence to suggest that the general public would like to see a greater provision of libraries, indoor sports centres, swimming pools, theatres and cinemas (Cosgrove and Jackson, 1972).

A further set of leisure activities requires the direct provision of open space. At a local level, public parks and recreation grounds are needed for a wide range of purposes: the principal activities pursued in these kinds of public open spaces include walking, picnicking, sunbathing, watching or participating in informal games or, in the case of children, merely playing together. But at a regional or national level, open land is needed for holiday excursions, golfing, horse-riding, racecourses and also simply for motoring for pleasure. Increased mobility, which goes hand in hand with economic development, leads to a considerable increase in travel for pleasure. Whereas some 80 000 visits were made to the Norfolk Broads in East Anglia in 1930, by 1970 the number had increased to 300 000 with people coming from further and further afield (Hall, 1966) and similarly, the number of visits to the Peak District National Park doubled from 4 million to 8 million in the 10-year period from 1962. Despite the obvious demand for the provision of areas like the National Parks or for the preservation of 'areas of outstanding natural beauty', it is probable that only a *very small percentage* of all visitors actually use the land so provided for any active outdoor pursuit, or bother to visit places of great beauty or interest that are to be found at some walking distance from the main roads through the area. The main reason for this is to be found in the fact that the public tend to derive more pleasure from a drive in the countryside than anything else. To some extent this pattern is rather reminiscent of Stephenson's suggestion that 'To travel hopefully is a better thing than to arrive' (Brancher, 1972). Blacksell (1971, p. 145) discovered in a survey of visits to the Dartmoor National Park that 'the majority of visitors come to the moor with only the most vague aims. Over three-quarters (76 per cent) of those interviewed gave as their main reason the wish for a quiet drive in the open country'. When they do eventually stop, most people are content either to sit in the car or to picnic or play close by: Wager (1964) found that 55 per cent of the visitors to Berkhamsted Common were either sitting in or picnicking near their cars; Burton (1966) found that 61 per cent of the visitors to Box Hill (Kent) were similarly occupied and Blacksell (1971, p. 146) found that 'most' visitors to Dartmoor 'agreed that they would park their car, picnic, sunbathe and perhaps have a short stroll'.

Patently, the provision of open countryside as an amenity is, in some ways, an exorbitant use of land in view of the apparent light use which is made of much of it. On the other hand, there can be no denying the fact that escape to the country (or even to jammed country roads for that matter!) is an essential amenity which must be increasingly provided, particularly since the amount of private individual space such as gardens that is provided in modern residential developments is very limited.

Further reductions in the length of the working week, together with further increases in income, paid holidays and personal mobility will be inevitable in the future. It has been projected that by 1985 the length of the working week in Britain will be only 75 per cent of its 1960 level, the average length of annual holiday will have doubled from its 1960 level and the number of privately owned cars will have almost quadrupled in the same 25-year period (Burton, 1967). The effect of such changes on demand for the provision of public open spaces of different kinds is bound to be considerable. In the USA, for example, it is estimated that by the year 2000 the number of recreation visits per head of population will have increased sixfold since the mid-1950s, consequent upon changes in living standards and potential mobility.

The increasing demands on land for agricultural, industrial and recreational purposes consequent upon increases in standards of living are further aggravated by the changes in population numbers and distribution that accompany economic development. The total world population has tended to increase at a generally exponential rate: it is estimated that there were about 250 million people in the world at the time of Christ, and that by the year 1500 there were twice that number; the number had doubled again by 1850 and again by 1940 when there were 2000 million people. During the next 30 years, another 2000 million people were added, so that by 1970 the total world population was sixteen times greater than its level at the beginning of the Christian era (figure 4.4). The present rate of world population growth is in the order of 2 per cent per annum and represents the difference between the present average birth rate of thirty-eight persons per 1000 population per year and the average death rate of eighteen persons per 1000 population per year.

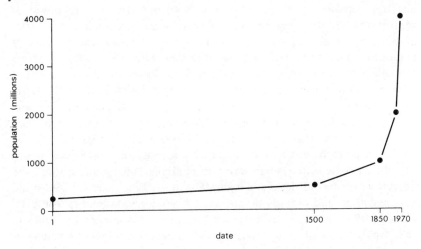

Figure 4.4 GROWTH OF WORLD POPULATION

The fundamental reason for this rapid increase in population is that death rates have been progressively lowered as increases in standards of living have led to better diets, medical knowledge and standards of social welfare, but birth rates have remained largely unaltered. It has been estimated that in Bronze Age times the death rate was as high as fifty per 1000 population per year and that the average length of life was in the order of 20 years. In the more advanced nations of the world, death rates have now been lowered to below ten per 1000 per year, and the length of life has increased to about 75 years. Naturally, the effectiveness and the extent of medical services, social welfare programmes and dietary levels is determined by the extent to which any society is aware of them or is able to adopt them, and it is not without significance that there are marked differences in death rates between social classes, which tend to reflect this situation. In France, for example, during 1970, the infant mortality rate was fifteen per 1000 in the upper classes, compared with fifty in the labouring classes—a difference that may be ascribed largely to social attitude, conditions of life and relative wealth, all of which are in turn reflections of different states of knowledge, educational background and achievement levels.

In comparison with death rates, birth rates have declined very little until recently, and yet fertility has generally been controlled from far earlier times than mortality. Even before the advent of modern contraceptives and other techniques for limiting births, a wide range of social customs had effectively 'controlled' fertility (Wrigley, 1967). In some societies, such as those of Western Europe before the Industrial Revolution, not only was the age of marriage considerably higher than it is now, but also the proportion of women who had never married was higher. In other societies, including many of those in the underdeveloped areas of the world today, the practices of polygamy, infanticide and abortion together with other social customs forbidding remarriage have had similar effects. In certain circumstances, as economic development begins, increased wealth coupled with changes in social structure and attitudes to life may actually cause birth rates to *increase*; having more real income and faced with increasing opportunities, individuals may begin to feel that they can afford to have larger families. However, as education proceeds this attitude is normally reversed since 'the development of education acts in several ways: it multiplies the preoccupation and attractions of the individual and it gives him at the same time, knowledge and new ambitions and changes his attitude to family responsibilities' (Beaujeu-Garnier, 1966, p. 131). An increasing awareness of responsibility either to self, family or social group, coupled with an increasing desire to achieve or succeed in life leads to the recognition of the need for limitation in family size, and progress in medical technology makes this need a practical possibility through the development of contraceptive devices, abortion and sterilisation techniques.

There are, of course, widespread differences between various classes of society and between locations in the extent to which these attitudes are adopted. Religious beliefs and other ethical codes may cause some groups to adopt different attitudes both towards children and towards birth control (Roman Catholic birth rates are generally about 5 per cent higher than Jewish birth rates and 11 per cent greater than Protestant birth rates), but the main cause of variation still appears to be educational background and motivational level. In 'developed' societies, for instance, the smallest families are normally those in which the desire to achieve is very great but that have not yet achieved all that they would like; conversely, the most prolific classes are those in which ambition is lacking and in which educational level is correspondingly low (McClelland, 1961). Similarly, the marked differences in fertility rates that exist between town and country regions, and even within urban areas (where birth rates increase in proportion to increased distance from the city centre) may reflect the fact that urban conditions of life make children more of a burden and that, in any case, the highest 'n-achievement' levels (section 2.3.2) are found in town dwellers.

The rate at which both birth rates and death rates are changed as economic and educational development proceed, determines the overall rate of population increase. In areas where technology, education, aspiration and prosperity are lacking, high birth rates (more than forty per 1000 population per annum) are matched by equally high death rates (more than thirty-five per 1000 per annum), so that the overall population growth rate is very small and usually less than ten per 1000 per annum. This *primitive type* of population is characteristic of most pre-development systems and indeed the pre-industrial societies of Europe were of this general form (figure 4.5A).

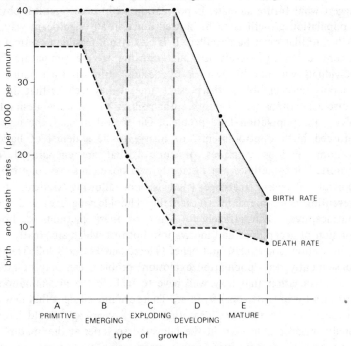

Figure 4.5 TYPES OF POPULATION GROWTH

As development proceeds, death rates begin to decline (to about twenty per 1000 per annum) for the reasons already given, without any change taking place in the high birth rate. In such instances, the overall growth rate is between twenty and twenty-five per 1000 per annum. This *emerging type* of population growth is characteristic of most African countries, the Middle East, eastern Asia and parts of Latin America (figure 4.5B). If the death rate is lowered even further (to about ten per 1000 per annum) while the birth rate remains largely unaltered, the natural growth rate increases very rapidly—often well beyond thirty per 1000 per annum—(figure 4.5C). This *exploding type* of growth is not, however, very common because more normally, as the death rates go below the ten per 1000 level, there is a reduction in birth rates to a level of about twenty-five per 1000 per annum and the overall growth is therefore brought down to a rate of between fifteen and twenty per 1000. This *developing type* is characteristic of Australia, New Zealand, Argentina, the USA and the USSR (figure 4.5D). Finally, a *mature* stage may be reached when both birth and death rates have fallen to a low level and the overall population growth has once more fallen to a rate of less than ten per 1000 per annum (figure 4.3E).

Because it is difficult to assess the extent to which education will bring about changed attitudes to population growth, it is equally difficult to suggest what future increases in population might occur, yet it is obvious that if population growth is to be slowed down in any significant way, it is the birth rate that must be curtailed. It is clearly not feasible to retreat from the present death rate levels without resorting to the premature killing of individuals whether by war or by design: although such action may be generally unacceptable at the present time, it may well be that the ethics of future generations would regard euthanasia as a normal and desirable way of solving the population–land problem. On the other hand, the suggestion of enforced birth control seems, to many, to be a denial of basic human freedom. Various measures of control that are encouraged, officially permitted or sometimes just discreetly overlooked in some countries, include abortion, delaying marriage (that is, not allowing students to marry, preventing child marriages, raising the school-leaving age) and the use of contraceptives. The efficiency of any of these methods is, however, a function of acceptance and motivation, both of which are greatly influenced by education, social class and belief (Tietze and Potter, 1962; Tietze, 1963). Consequently, the 'population explosion' problem can only be attacked by greater education that may well have to include the dissemination of new attitudes towards personal freedom. There are indeed signs that new attitudes are beginning to emerge: ideas that only 5 years ago would have seemed totally unacceptable even in the educated societies at the 'mature' stage of development, are now being rather more seriously contemplated. Thus the utopian suggestion that the introduction of certain additives to water supplies to cause temporary sterilisation of the individual, which could be overcome— if children were *really* desired—by taking a pill to re-establish temporary potential fertility, may well become reality if attitudes continue to change at their present rates.

By 1972, there was some evidence to suggest that birth rates in many nations (Britain included) had been reduced to such an extent over the preceding 5 years that if the rate of decrease was maintained, the population growth rate would reach 'replacement' level within 5 years. It is difficult to be sure whether changes such as these represent continuing trends or whether they are temporary phases out of line with other trends in either birth or death rates. For this reason, estimates of future population growth rates are difficult to make, and vary according to the assumptions that are made about birth and death rate trends; the lowest estimate puts the total world population at 5500 million by the year 2000, the highest at 7000 million and a 'medium' estimate is for 6000 million. Whichever figure is chosen it is apparent that further demand for land is bound to arise in the future, though the pressure will vary from location to location because as population continues to increase and as standards of living improve, changes also begin to occur in the distribution of population within nations and regions.

The operation of scale economies in the process of spatial organisation is such that as economic development proceeds, industry and population begin to agglomerate (sections 3.2 and 7.3.3) thereby causing the population of urban areas to increase relative to the population of rural areas (table 4.2). Despite this tendency, which is, of course, most marked in the 'developing' and 'mature' regions of the world, six out of ten of the world's population

TABLE 4.2 URBANISATION OF POPULATION: percentage of population living in urban areas, 1910-70.

Country	1910	1930	1950	1970
Australia	–	36.4	49.3	67.8
Bulgaria	19.1	31.4	24.8	31.2
Ecuador	–	25.1	31.0	36.7
France	44.2	51.2	52.0	58.9
Germany	61.7	69.8	70.8	76.4
Japan	19.0	24.6	52.0	69.8
Spain	34.8	42.6	56.7	64.7
Sweden	24.8	38.4	52.3	62.4
UK	52.3	59.8	64.7	71.7
USA	45.7	56.2	60.7	69.7

still live in rural surroundings, so that in many ways it can be argued that 'this is a rural world in spite of the fact that urbanisation is more advanced than it has ever been' (Jones, 1966, p. 13). On the other hand, there can be little doubt that this is merely a transitional phase that will be over by the turn of the century when urbanisation will not only have occurred more extensively in the currently 'emerging' regions but will also have been intensified in the 'mature' regions. (Most estimates suggest that the proportion of people throughout the world living in cities of greater than 20 000 population will have reached 45 per cent by the year 2000 and 90 per cent by 2050.)

As agglomeration proceeds, variations in the geographical pattern of population density around and within urban areas begin to emerge. Not surprisingly, density tends to decrease as distance away from the centre of urban areas increases, the relationship being negatively exponential and of the general form

$$P_d = P_o e^{-bd}$$

where
P_d = population density at a given distance d from the centre
b = the slope of the density decline curve
P_o = the density of the central area *extrapolated* from the slope for outer areas (Clark, 1951).

The *density lapse rate* has been identified in several different areas and is found to characterise both the regional pattern of population density *around* nodal urban areas and the pattern *within* urban areas (figure 4.6A-B). Differences in the density lapse rate within urban areas are also distinguishable between locations along major routeways into the centre and locations in the interstitial areas: densities are normally higher along the routeways than in the areas between them (figure 4.6B).

Through time, the expansion of urban areas and the progressive intensification of residential development cause the lapse rate to change. In most cities of the Western world, the lapse rate has steadily decreased since the late nineteenth century, though more recently its *rate* of decrease has been accelerated as higher-density development has occurred near the periphery and as more and more people have moved out from zones near the centre of town (figure 4.6C; see also section 7.3.3). In other regions of the world, however, density appears to have increased at a constant rate in all parts of the city (Berry, Simmons and Tennant, 1963; figure 4.6D).

Changes in the distributional pattern of population together with the projected increases in population numbers place further demands on the land resources of the system and necessitate a careful evaluation of alternative sites in the light of the additional demands created by increases in standards of living. The fundamental cause of the land scarcity dilemma is consequently

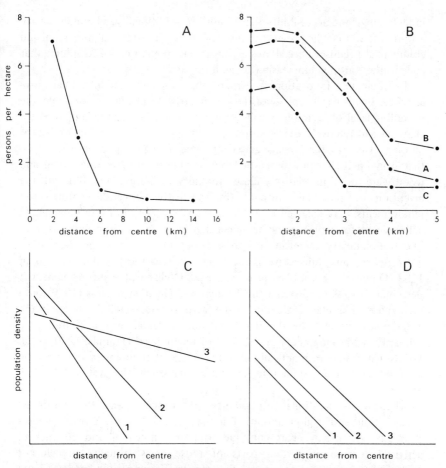

Figure 4.6 POPULATION-DENSITY LAPSE RATES A—density and distance around a nodal point (Exeter region 1972); B—density and distance within the urban area (line A represents the mean condition; line B the lapse rate along major routeways into the centre; line C, the lapse rate in the interstitial areas) (City of Exeter, 1972); C—changes through time in western cities (line 1 represents the earliest pattern, lines 2 and 3 represent successive periods); D—changes through time in other regions.

the phenomenon of economic growth and development that is, in turn, one of the central motivations of the whole system. Unless this motivation is revalued, there can be little hope of alleviating the problem.

4.1.2 *Obsolescence and decay*

The corollary of growth is obsolescence since growth, of necessity, entails a revaluation of all aspects of spatial organisation consequent upon the changes in technology and accessibility that it induces. Additionally, all buildings have a finite life so that the progressive physical decay of residential, industrial and commercial premises necessitates a continuous programme of rebuilding that adds to the continued increasing demand for land.

Technological innovation is perhaps the root cause of locational obsolescence. In the earliest phase of the iron industry in Britain, for example, the prevailing level of technology called for easily-mined ore, abundant wood for charcoal and relatively easy transport; smelting took place on the Catalan (or bloomery) hearth that was operated by a pair of hand-operated bellows, so that iron smelting could be carried out at almost any site where iron and wood were found in relatively close proximity to the markets. But with the invention of the blast furnace (in the early sixteenth century) site requirements were changed as water power was needed both for blowing the bellows and for working the hammers that beat pig-iron into wrought iron. The iron industry therefore had to seek out sites in the upper reaches of powerful streams. Successive innovations, such as the puddling process in 1783 (Onions), the hot-blast process in 1828 (Neilson), the steam engine, the Bessemer process (1856) and the Thomas and Gilchrist process (1879) have each caused a similar revaluation of the factor of site, so that 'a change in the geographical pattern of the iron and steel industry was a consequence primarily of developments in the field of technology' (Pounds, 1959, p. 7). Locational change in practically all other forms of economic activity can be seen to result from similar technological changes in the system (Mitchell, 1954).

Changes in locational accessibility often accompany the continued urbanisation and agglomeration of human activity as the growth process proceeds. As urban areas continue to grow, industrial and commercial activities located at certain parts of those areas (usually, though not exclusively, at the centre) begin to find that traffic congestion becomes acute or that suppliers and customers alike have moved to new locations (often on the periphery of urban areas; section 7.3.3). These conditions may reach such proportions that the activities concerned may seek new premises at locations that appear to possess greater relative accessibility in the newly emerging locational system. A form of ECONOMIC BLIGHT may then characterise the areas in which accessibility has decreased—the outward manifestations of which are the large number of vacant premises which arise. Accessibility may also be revised through the operation of scale economies that create the need for many activities to increase their scale of operation (section 3.1). For many firms, there simply may not be any land available for extending their present premises and the result may be that the firms migrate to a larger site that has room for future expansion. FUNCTIONAL BLIGHT of this kind (Berry, Tennant, Garner and Simmons, 1963, p. 180) may have the same *effect* as economic blight though the *cause* is rather different. The tendency for retail stores to migrate within the CBD to successively larger premises is a common form of this kind of obsolescence.

Similar locational changes may be necessitated by the gradual physical decay of buildings that is 'a continuous process which begins as soon as a building has been completed and increases as the building ages' (Berry, Tennant, Garner and Simmons, 1963, p. 195). Many factors affect the rate at which buildings become 'blighted' in this way, of which the most significant are the type and standard of construction and the extent to which successive occupants have maintained the property.

Physical decay is notoriously difficult to define or measure and is often rather subjectively assessed in terms of whether the building is in 'good, fair, poor or very poor' condition. Attempts have been made to give greater precision to this scheme of grading by the use of 'scores', though the process still remains largely subjective (table 4.3; Medhurst and Parry-Lewis, 1969).

TABLE 4.3 PHYSICAL BLIGHT: indices and scores suggested by Medhurst and Parry-Lewis (1969). Scores ranked according to relative cost of rectifying listed faults.

Signs	Extent			
	much	some	little	none
Surface deterioration	5	3	1	0
Paint peeling	3	2	1	0
Displaced roof units	9	5	1	0
Broken glazing	7	3	1	0
Gutter leak	7	3	1	0
Settlement (cracks)	11	6	3	0
Timber rot	8	4	2	0
Sagging roof	10	6	2	0

The scale of the problem clearly varies from region to region depending on the proportions of buildings of particular ages, but it is perhaps surprising that in Britain it was estimated that in 1966 12 per cent of all housing was physically unfit for habitation, that a further 28 per cent had less than 30 years life expectancy and that some 15 per cent of all industrial and commercial premises had a life expectancy of less than 25 years.

As standards of living increase and as technological innovation leads to the introduction of new amenities, some buildings, particularly residential ones, will become obsolete in the sense that they fall below the accepted standards of amenity provision. The provision of such amenities as running hot and cold water, WCs and fixed baths is one of the main advances in twentieth-century

TABLE 4.4 AMENITY PROVISION: percentage of all residential buildings lacking certain facilities, USA and UK 1965.

Facility lacking	USA	UK
Running cold water	7.1	0.7
Running hot water	12.8	16.6
WC of any kind	10.3	4.8
Internal WC	3.0	13.4
Fixed bath	14.8	18.5

housing conditions, but there are many dwellings (mainly of some age) that still do not have these basic services (table 4.4).

Physical decay and obsolescence of these kinds can be treated in two ways: either buildings can be brought up to standard, or they can be demolished. Which course of action is taken largely depends on the relative costs involved. In fact, the provision of the basic services may not be too expensive (consultants' estimates suggest that it costs £80 to provide a WC in an existing room, or about £450 if a new room must be added), but if the dwelling has a low physical expectancy of life, then the expenditure may not be justified. Maintenance and physical refurbishment costs are highest for buildings in a poor condition so that it may be thought more cost-effective in the long run to replace dwellings in poor condition by new ones. On the other hand, if dwellings are in fair or fair–poor condition and yet have relatively high age expectancy, it may be cheaper (or better) to refurbish them. The discount factor is thus an important consideration in the establishment of policies towards decay.

The standard of the physical environment in which houses are located has also to be considered in assessing the scale of the problem of renewal and

TABLE 4.5 ENVIRONMENTAL BLIGHT: indices and scores suggested by Medhurst and Parry-Lewis (1969).

Signs	Extent			
	much	some	little	none
Offensive smells	3	2	1	0
Air pollution	3	2	1	0
Noise	3	2	1	0
Grass/trees	0	1	2	3
Litter	3	2	1	0
Parked vehicles	3	2	1	0

blight. The environmental conditions of residential areas can be estimated by similar methods to those employed in the estimation of physical building blight (table 4.5); on this basis it is estimated that about 13 per cent of all residential areas in Britain are already in poor condition (Stone, 1970). Most of those areas could only be improved by demolition—a remedy that creates still further demands for new residential, industrial or commercial development.

4.1.3 *Intensity*

It becomes increasingly clear that, in the absence of a reduction in growth, obsolescence and decay, the only way in which the increasing demands for residential, agricultural, industrial, commercial and recreational land can be partly met is through the intensification of land use. Increases in land productivity can, however, only be made through the substitution of capital (in the form of technological innovation) for land, and to that extent it is surely paradoxical that the very mechanism that appears to create the land dilemma itself appears to hold the key to its solution. If this were the case, growth would constitute a self-regulating mechanism in the landscape system and the problem would ultimately be resolved through the application of successively more and more advanced technologies. Yet it is probable that there are definite *limits* to the possibilities of capital–land substitution, even if those limits may not yet be fully realised, and it is this probability that has led many 'prophets of doom' from Malthus (1798) through to the neo-Malthusian latter-day Adventists to consider the growth mechanism as a positive feedback relationship in the system.

Instances of the possibilities of land use intensification can be found in all parts of the system. The amount of land demanded for agricultural production has in many instances actually decreased because improved technology has made possible considerable increases in crop yields (table 4.6) and indeed, it has been estimated that in the USA, 'the total amount of cropland "saved" through yield increases in the past two decades is perhaps 200 million acres' (Landsberg, 1963, p. 13). For the future, the introduction of various factory farming methods (such as battery beef, pork, hens and eggs) should lead to a further reduction in the hitherto extensive demands for land made by most forms of agricultural activity. The yield increases that may be made possible will vary from crop to crop, as they have in the past, but the net result of such increases, considered together with anticipated future developments of agricultural substitutes, is that the amount of land needed for agricultural activity by the year 2000 is estimated to be less than that actually used in 1956.

The effects of capital substitution in the industrial sector have been

TABLE 4.6 CROP YIELDS: changes in yields (100 kg/hectare) between 1948–52 and 1968. Source: F.A.O. Yearbook (1970).

Commodity	World			Europe			N. America			Africa		
	1948–52	1968	% change	1948–52	1968	% change	1948–52	1968	% change	1948–52	1968	% change
Wheat	9.9	14.6	47.4	14.7	25.3	72.1	11.6	17.7	52.5	6.0	9.2	53.3
Rye	9.6	15.0	56.2	14.6	21.4	46.5	7.9	13.5	70.8	4.1	3.9	−4.8
Barley	11.3	17.4	53.9	16.9	29.7	75.7	14.5	21.7	49.6	6.8	10.1	48.5
Oats	11.4	16.8	47.3	16.0	23.7	48.1	12.7	18.9	48.8	7.0	6.1	−12.8
Maize	15.9	23.7	85.1	12.4	29.2	139.3	24.9	49.4	98.3	8.0	9.8	22.5
Millet	5.1	7.7	50.9	9.7	27.3	181.4	12.6	33.2	163.4	5.4	6.4	18.5
Rice	16.3	21.5	31.9	43.0	41.5	−3.4	25.6	49.6	93.7	9.8	14.1	43.8
Potatoes	110.0	138.0	25.4	137.0	187.0	36.4	155.0	231.0	85.0	51.0	60.0	17.6

similarly considerable. In particular, industrial demands for natural resources have been relatively reduced through increased efficiency in their use. Aggregate efficiency in the use of energy resources is estimated to have increased from 10.5 per cent in 1860 to 22 per cent in 1950 (Putnam, 1954). In 1965, research workers at America's 'Resources for the Future' claimed that 'the pound of coal that was burned up in the course of generating 1 kilowatt-hour of electricity 20 years ago would produce 1.5 kilowatt-hours today. At the 1960 rate of 700 billion kilowatt-hours generated thermally, the savings in fuel consumption due only to this 20-year progress in efficiency is the equivalent of some 165 million tons of coal' (Landsberg, 1963, p. 13).

The actual total space requirements of industrial premises has generally increased as firms have attempted to derive scale economies of production, yet in twelve American industries studied by Osborn (1953), space requirements *per unit of output* were reduced on average by 58.8 per cent consequent on the introduction of automated processes. In some instances, automation may even lead to a *net reduction* in the actual space required for the operation of a particular process. Such at least appears to be the case of office and administrative processes based on computer operation, since even if the central office of a particular firm may require more space to house the computer, regional sub-offices should require considerably less.

Improved building technology can also increase the effective 'yield' of land by the perfection of techniques for high-rise building. Savings of residential land in particular can be effected by the construction of such blocks, though the actual cost of high density high-rise development is *greater* than the traditional form of one or two-storey buildings. In general, it costs about half as much again to house a group of people in fifteen-storey blocks as in two-storey housing. The explanation of this is relatively simple: less land is needed for each person housed in high-rise flats, but the land that is saved would be worth only about £200 per acre if it were to be used agriculturally. However, the real cost of providing and maintaining dwellings rises more rapidly than the need for land falls so that the net cost of saving land increases. In other words, 'building high is an extravagant way of saving land' (Stone, 1970, p. 157).

The continuing process of agglomeration is, in itself, one of the mechanisms by which all activities are able to reduce their demands for land since intensification is synonymous with agglomeration. Residential densities are always highest in agglomerated areas, probably because there is every incentive to minimise the transfer costs that would be incurred if residential development were to sprawl further and further away from the industrial and commercial centres. Savings of land can also be made in larger settlements for commercial, industrial and other uses such as streets and footpaths (Bartholomew, 1955). The number of industrial businesses and plants shows a close accordance with the size and distribution of urban areas: in general, as the population of urban areas increases the number of businesses and plants also increases (figure 4.7) although the increase is not necessarily in direct *pro rata* proportion because of the tendency for plants and businesses to expand their premises with increased demand rather than for new plants and businesses to be created. The amount of land required for industrial and

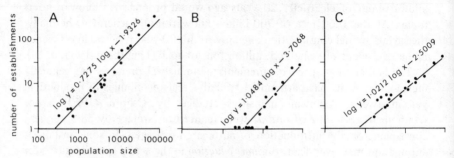

Figure 4.7 CORRELATION BETWEEN BUSINESSES AND SIZE OF SETTLEMENTS
A—chemists' shops; B—furniture stores; C—grocery stores (Exeter region, 1969).

commercial premises is therefore *proportionately* less the larger the agglomeration (figure 4.8). Similar proportional savings can also be made in the provision of roads, railways, streets and footpaths: the amount of land needed per head of population for these services is actually less in large conurbations than in small towns (figure 4.8; Haggett and Chorley, 1969; Owens, 1968).

Despite these obvious possibilities there are some problems and limitations involved in land use intensification. Ecological conditions of soil set obvious limits on agricultural productivity and it is for this very reason that the relationship between ecological and man-made systems is, as was suggested earlier, so intricate and delicate. High-rise development is also limited by physical and social conditions: in order to build high it is necessary to use relatively flatter land than may otherwise be the case and greater attention has to be paid to daylight requirements. The higher the buildings, the further apart they must be sited in order to provide sufficient daylight for the residents. In high latitudes (such as Britain), tower blocks have to be more widely spaced than in lower latitudes (such as New York) simply because of the greater lengths of shadows cast by the midday sun in winter (figure 4.9). Alternative designs have to be developed in order to circumvent the problem and it can be demonstrated, in theory at least, that 'much greater densities can be achieved by the use of broken cruciform layouts, by reducing the size of rooms, or by making the blocks sufficiently deep to enable dwellings to be placed either side of a central corridor or back to back' (Stone, 1970, p. 105).

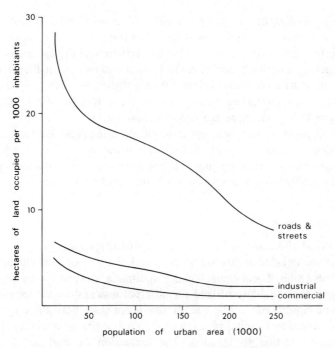

Figure 4.8 SPACE REQUIRED FOR SERVICES IN URBAN AREAS Nord and Pas de Calais Départements, France (1965).

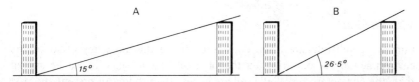

Figure 4.9 SPACING OF HIGH-RISE DEVELOPMENT A–London; B–New York.

In housing areas of dwellings with their own private gardens, density is mainly determined by the size of garden, and indeed in modern developments (especially in the public sector which provides housing for low income groups), it is by cutting down on garden size that considerable land economies can be achieved. Whether this method of increasing densities, and consequently of reducing demand for housing land, is desirable in terms of amenity is, however, open for debate, especially in view of the importance that so many people attach to gardening (section 4.1.1). Sociologically and psychologically too, it can be argued that high-rise living is detrimental both to the individual and to society as a whole. In blocks of flats, community feeling is frequently non-existent (a fact that many would argue is a loss to society), hooliganism is more evident and psychological disorders more common than in traditional semi-detached or terraced housing estates.

Life in urban agglomerations also has certain sociological, psychological and practical drawbacks which could be regarded as potential limitations to this form of land economy. Crime rates are higher, neurosis is more evident, congestion and travel-time wasting is more acute and, beyond a critical size, life begins to become more expensive (section 3.2.2).

For all these reasons, land use intensification remains a limited possibility and indeed, because some of the side effects are so apparently undesirable, many governments have actually imposed constraints on such development in order to establish minimum reasonable living standards (chapter 9).

4.2 EVALUATION

The optimal allocation of land between conflicting uses can only be achieved through the detailed evaluation of the physical and economic characteristics of available sites. Local characteristics of climate, soil and topography partly determine the suitability of land for different uses at different locations, but economic conditions may be equally significant since both the price of land and the amount of capital investment required to change or develop land use vary from location to location. The evaluation of land for residential, agricultural, industrial, transportational or recreational purposes is thus based on the assessment of both physical and economic conditions.

4.2.1 *Physical*

Climatic conditions are often considered to be the most important of the environmental factors affecting the physical system since it can be argued that 'the characteristics of the soil are largely the product of present and past climates and the vegetation that has flourished in them, and the effects of relief are to no small degree expressed through resulting climatic variation' (Symons, 1967, p. 21). The local 'microclimate' is, in this respect, probably even more important than the overall 'macroclimate' of an area, since important climatic modifications can be caused by local topographic conditions (Geiger, 1950). There are considerable differences between locations in valleys, on valley slopes and on the tops of valleys under different climatic regimes in terms of temperature and humidity conditions. On an anticyclonic day in spring, for instance, the range of air temperature is greater in the bottom of a valley than it is on the slopes or at the top and nocturnal temperature inversions may also be established (figure 4.10A). The daily course of relative humidity follows a similar pattern (figure 4.10B) and, with certain combinations of temperature and humidity, early morning frost or fog

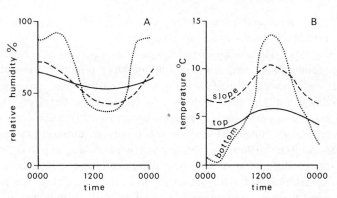

Figure 4.10 MICROCLIMATIC VARIATIONS A—changes in relative humidity by time and location (Exe Valley, April 14, 1969); B—changes in temperature by time and location (Exe Valley, April 14, 1969).

'hollows' may form in the bottom of valleys. The 'aspect' of different slopes is also fundamental in affecting temperature intensity: north facing slopes are always cooler than south facing slopes and the degree of slope may also cause variations in the location of highest temperatures. Such climatic variations may be significant in the evaluation of alternative land uses at different locations: it may, for instance, be necessary to avoid the use of certain valley bottom sites because of their liability to frost or local fog; similarly, the problem of smoke or fume disposal from industrial premises may necessitate a choice of site where local microclimatic conditions ensure the rapid dispersal of fumes into the atmosphere without affecting any nearby residential or recreational areas. The spatial arrangement of land development may also be affected by local climatic conditions since buildings themselves create their own microclimates: in particular, the physical arrangement of buildings of different heights separated by different widths can, under certain circumstances, create wind-tunnel or vortex effects which must be avoided if at all possible. Recreational land must equally be chosen with respect to its liability to fog, frost, wind or even adverse heat conditions.

Soil conditions are related to variations in soil texture, drainage, temperature, infiltration and fertility. Texture is determined by the proportions of sand, silt and clay that are found in soils at different locations, and texture in turn conditions the *latent* fertility of soil. Clay soils are potentially the most fertile even though they are essentially 'heavy' in nature; loamy soils are often regarded as ideal because they are the easiest to work and can be used for almost any kind of agricultural activity without loss of yields; whereas sandy soils are basically infertile since they are chemically inert. Variations in the combinations of sand, silt and clay fractions lead to associated variations in potential fertility: a sandy loam is potentially more fertile than a loamy sand because its silt and clay fractions are greater, but it is not so potentially fertile as a sand–clay loam or any other soil whose clay fraction is greater. The acidity of the soil also has a bearing on the potential range of commodities that the soil can produce: potatoes, for instance, require a slightly acid soil (pH between 5.5 and 6.0) whereas barley prefers more alkaline conditions (pH of about 6.5). The potential fertility of the soil is additionally related to the cation-exchange capacity (the capacity of soil colloids to absorb nutrients) and to the supply of nutrients in the soil (Buckman and Brady, 1960), but soil temperature may also help to define the kind of commodities that a particular soil can support. Soil temperature, however, is determined by the microclimate of site, the tilth of the ground and the type, texture, depth and moisture of the soil, so that climatic, soil, and topographic conditions are all mutually interrelated. Topography indirectly causes modifications to climatic, soil and vegetational conditions: temperatures generally decrease and rainfall and wind correspondingly increase with increased altitude so that soil and vegetational conditions also vary according to altitude. Crop yields respond directly to variations in these conditions. The best yields of wheat, for example, are found in areas of flat topography with good, deeply worked, fertilised soil that has had an enriching previous crop and where there is no hot spell but a moderate amount of rain in the period from earing to harvest; as any of these conditions worsen in different combinations, yields progressively diminish (figure 4.11; Azzi, 1958).

Topographical conditions alone may also contribute to the suitability of different sites for different uses, though the effect may vary from location to location. In a temperate location such as Britain, upland areas become 'marginal' not only for agricultural activity but also for residential, industrial and, possibly, recreational activities; whereas in tropical locations upland conditions may be more 'moderate' than lowland conditions and, consequently, many forms of activity seek upland sites. Slope conditions may directly influence the siting of any development through the various cost and other limitations that they tend to impose. It is normally easiest and cheapest to build on relatively flat sites but it is often argued that the best sites for residential development are those of 'gentle' slopes, that allow architects and planners to diversify the site and that afford interesting views for the residents. Improvements in building technology have successively made development on steeper and steeper slopes feasible, though as slopes increase so does the cost of building, since more extensive foundations and reinforcing of the building are necessitated; similarly, improved technology has made possible the use of steeper and steeper slopes for agricultural purposes, though many farm implements are still limited to use on slopes of certain gradients (figure 4.12; Curtis, Doornkamp and Gregory, 1965).

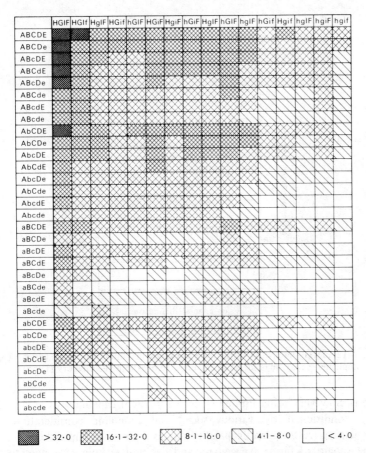

Figure 4.11 ECOLOGICAL CONDITIONS AND CROP YIELDS Yields on Mentana wheat in quintals per hectare (1953–4) under different combinations of ecological condition. *Key* A, good soil; B, flat topography; C, deeply worked soil; D, fertilised soil; E, enriching previous crop; F, sufficient rain in the month before earing; G, no trace of rust; H, moderate rain earing to harvest; I, no hot spell; a, poor soil; b, very hilly topography; c, shallow worked soil; d, not fertilised; e, exhausting previous crop; f, drought in month before earing; g, attacks of rust; h, excess rain earing to harvest; i, hot spell. Source: Azzi (1958).

The physical evaluation of land for different purposes is thus normally based on assessments of variations in climate, soil, site and topography though, in detail, the emphasis placed on each of those conditions varies according to the purpose of the evaluation. Many different systems of land classification for agricultural purposes, for instance, have been proposed for different regions (Symons, 1967, pp. 239–57; Morgan and Munton, 1971, pp. 122–5). The usefulness of classifications based on physical characteristics

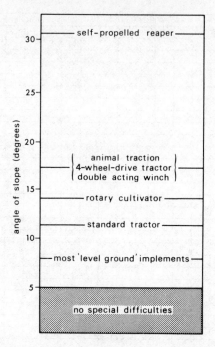

Figure 4.12 LIMITATIONS ON MECHANISATION maximum slopes on which various implements may be used. Source: Curtis, Doornkamp and Gregory (1965).

alone is, however, rather limited since the inherent capability of land is also partly conditioned by various social and economic considerations. In particular, agricultural land capability relates as much to farmers' capabilities in the sphere of farm management and decision making as it does to the physical ability of land to support any form of agricultural activity. For this reason, one of the assumptions usually made in land capability classifications is that a moderately high level of farm management will prevail. Thus, the Soil Conservation Service of the United States Department of Agriculture has established eight land-capability classes on the basis of soil, site, climate and management techniques (Klingebiel and Montgomery, 1961). The Soil Survey of Britain has adapted this classification for Britain, and identifies seven classes of land capability on the assumption that 'land is assessed on its capability under a moderately high level of management' (Bibby and Mackney, 1969, p. 12; also table 4.7).

The evaluation of the scenic resources of any region, which is essential for the identification of areas which can be reserved for recreational purposes, is fundamentally based on variations in topographic conditions. This is because the individual's perception of natural landscape appeal is generally thought to

TABLE 4.7 LAND CAPABILITY CLASSIFICATION: Based on Bibby and Mackney (1969).

	Class 1	Class 2	Class 3	Class 4	Class 5	Class 6	Class 7
Gradient (degree of slope)	<3–7	<7	<11	<15	<25	level to >25	level to >25
Height (metres)	<152.50	<228.75	<381.25	<457.50	<533.75	<610.00	>610.00 (533.75 in west and central Highlands)
Mean daily temp (°C) April–September	>15	>15	>14	>12	>12	>12	>12
Rainfall less potential evaporation (mm)	<100	<100	<300	<600	<600	<600	<600
Special rainfall conditions	—	—	land over 122m must have <1016mm annual rainfall	<1270mm annual rainfall	land over 183m must have <1270mm annual rainfall	land over 305m must have <1524mm annual rainfall	—
Soil texture	Loams, sandy or silt loams, or humose variants deeper than 762mm or peat; stoneless or slightly stony	rooting depth >508mm stoneless or slightly stony possibility of wind erosion in eastern areas	rooting depth >254mm; textures vary from stoneless to stony and from sandy to clayey	soils sufficiently deep to allow ploughing; a wide range of textures	only rocky boulder-strewn soils which prevent mechanised improvement are excluded; mountain soils liable to severe erosion	extremely stony, rocky or boulder-strewn; severe erosion hazard on steep slopes	as class 6
Soil and land drainage	well or moderately well drained	moderately well or imperfectly drained	imperfectly or poorly drained	poorly drained requiring a comprehensive drainage scheme; occasional flooding (1 in 5 years)	very poor or poorly drained; subject to flooding (1 in 3 years)	very poorly drained peat or humose soils of uplands and estuarine marshes or undrained peats of lowland areas	very poorly drained

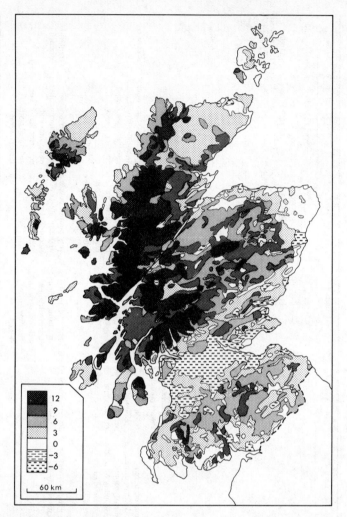

Figure 4.13 LANDSCAPES OF SCOTLAND Scenic resources assessed by point scores ranging from 12+ to −6. Source: Linton (1968).

be based on his assessment of relief characteristics as defined by the diffuse–dense, up–down, left–right, vertical–horizontal and delineated–open aspects of spatial perception (section 2.3.3). Linton (1968), for instance, suggested that a 'points score' system could be used to evaluate the scenery of any location: six 'landform assemblages' were identified, ranging from lowland areas that scored nought points to mountainous regions with a relative relief greater than 2000 feet (610 metres) that scored six points. Landscape beauty is not only related to topography, however: the land use of

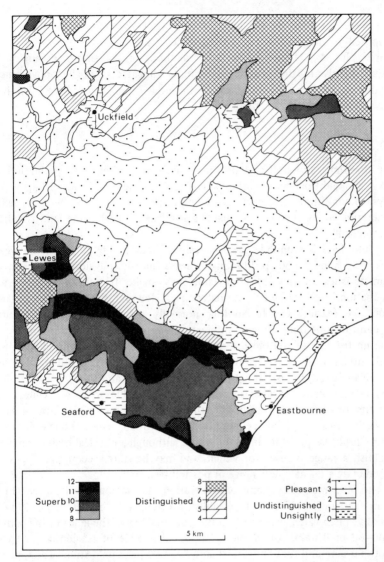

Figure 4.14 LANDSCAPES OF SE ENGLAND A point score evaluation on a scale ranging from 0 (unsightly) to 32 (spectacular). (In Britain, the range never exceeds 18.) Source: Fines (1968).

any area may be just as significant because certain uses of land may be regarded as unattractive and may detract from the 'natural' beauty that may have existed in the absence of man's occupance of a particular site. Linton therefore added a further series of points in his landscape evaluation scheme that were accorded on the basis of land use (urbanised areas, which were supposed to be unattractive, scored minus five points whereas 'wild' areas scored plus six) (figure 4.13). Other methods of assessing landscape value differ slightly in the criteria and points allocated, though topographical conditions are normally the basis of the evaluation (Fines, 1968; Penning-Rowsell and Harly, 1973; figure 4.14).

4.2.2 *Economic*

The economic appraisal of land must always accompany any evaluation of physical condition because the value of land varies partly according to its physical conditions and partly according to its location. In either case, value is determined by the decision maker's assessment of what the land is worth economically.

While 'land is immobile geographically, it may be highly mobile between uses' (Chisholm, 1966, p. 117) in the sense that the physical and locational characteristics of any plot of land may make it equally suitable for occupation by a number of alternative uses. However, the conversion of land from one feasible use to another normally requires considerable capital expenditure not only directly in creating the new use but also indirectly in clearing the site of the old use. Rural land may thus be converted to urban land, urban land to recreational land (as in the former CBDs of some American cities), industrial land to residential land, and so on, provided that it is economically feasible or even legally allowable so to do. Unless the benefits or the returns outweigh the costs involved, conversion of land between uses may be economically unjustified even though it may theoretically be possible. In other words, although a physical evaluation may establish a range of uses for which land may be suited, economic appraisal may dictate a far narrower range of possibilities.

The fundamental economic consideration of land concerns its price which is, in fact, rather difficult to establish precisely (Carter, 1972, pp. 194-6). Initially, price varies according to the potential use of the land: in the period 1960–62 in Britain, for instance, 'the median price of residential land was £800 per acre and for commercial land £25 000 per acre' (Stone, 1965, p. 3). But there are also marked regional and local variations in land values that reflect the influence of distance, density of development and considerations of accessibility generally. In Britain, regional variations in the price per acre of residential land are quite marked: the highest values are found in the London region and the Birmingham conurbation, but most land in urban areas is usually higher in value than land in the surrounding rural districts. As distance from the centre of the high land value area increases, so values tend to decrease. At 16 kilometres from the centre of London, land prices are 26 per cent lower than at the centre, and at a distance of 64 kilometres they are only 25 per cent of the central prices; in the Birmingham region a similar general relationship also holds, but the actual rate of distance-decay is rather greater (figure 4.15). One of the factors responsible for this difference in the

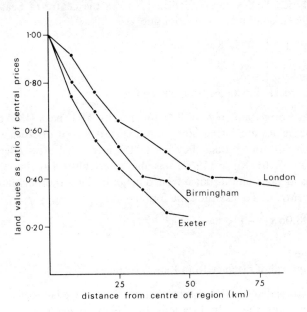

Figure 4.15 LAND-VALUES LAPSE RATES.

rate of decline is that of density of development. The greater the density of development, the greater is the price per acre demanded. Hence, values in the London region are consistently higher than in the Birmingham region at all distances, partly because of the greater density of development in the London region. Stone (1964) has shown that while the relationship between prices (per unit area) and density is essentially linear in form, that between price (per unit area) and distance is exponential, and the general relationship between price, distance and density takes the general form

$$y = \frac{1000}{a} - \frac{z}{b}(c + dx)$$

where
  y = price
  z = distance from centre of region
  x = density

and a, b, c and d are constants.

At the local level there are distinct variations in land values particularly within urban areas. In general, land in the city centre is more expensive than land at the city edge so that land values tend to decline with distance away from the city centre. Seyfried (1963), for example, found correlations of +0.64 between land values and distance from the centre of Seattle, while Knoss (1962) found that the relationship

$$y = 1691.1 + 19\,975.78\,x'$$

where
  y = land value
  x' = reciprocal of distance from city centre

described a correlation of $r = 0.799$ for the city of Topeka (USA). In most cities, land values are higher along the major roads out of cities and at the intersections of routes than they are in the interstitial areas (Berry et al., 1963; figure 4.16). Knoss (1962) was able to establish that the relationship between land value and distance from a major business thoroughfare running from the centre of Topeka took the form

$$y = 8498.06\,x_2' - 1182.5$$

where
  y = value
  x_2' = reciprocal of distance from Kansas Avenue

and was correlated to the extent that $r = 0.68$. Arterial roads and intersections are not the only locations that help to determine the relative accessibility of any plot of land, and it is clear that distance away from such facilities as shopping centres, bus stops, railway or tube stations and parks, together with considerations of social condition, may be equally significant in determining the price of land at any location. Thus Yeates (1965) was able to show that in the years 1910 and 1920 (before the development of major radial routeways) land values (V') in Chicago were related not only to distance from the CBD (C') but also to distance from regional level shopping centres (R'), subway system lines (E') and from Lake Michigan (M'), as well as to population density (P') and the proportion of non-whites at every location (N'), according to the general form

$$V' = a + b_1 C' + b_2 R' + b_3 M' + b_4 E' + b_5 P' + b_6 N' + e$$

Within the CBD itself, land values decline away from the peak value point that is usually associated with buildings located nearest to the hub of economic life in the district. In fact, land values decrease *very* rapidly in the area immediately surrounding the peak value point, but then the decline becomes less and less marked particularly towards the edge of the CBD. The peak values are limited, therefore, to a very circumscribed geographical area:

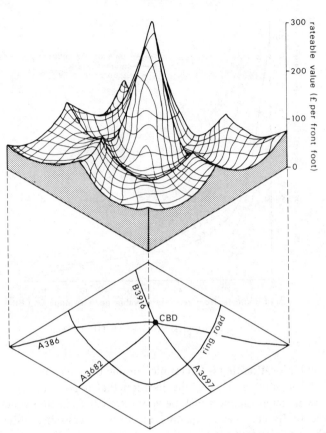

Figure 4.16 URBAN LAND-VALUE SURFACE hypothetical city (notice how land values are highest at the centre of the town, along the major roads into the centre, and at the points where the ring road intersects the main roads).

indeed, only 8 per cent of the total land contained within the CBD is normally valued at as much as a half of the peak value (Murphy and Vance, 1955). The rate of decline is also markedly greater in the 'zone of discard' than in the 'zone of assimilation' where new developments tend to force up the price of land (figure 4.17).

Land prices, like any others, are largely determined by demand and supply conditions so that land at any location is normally bought by the buyer who can pay the most for it with respect to all other potential buyers. But buyers are only prepared to bid a price for any site that is consistent with the advantage that they hope to derive from possession of that site. The 'accessibility' of any plot of land may in this way influence the use to which it may be put. Accessibility can be defined according to several different

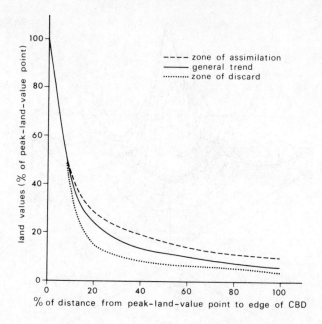

Figure 4.17 LAND VALUES IN THE CBD.

criteria and a location that is 'accessible' for one thing may not necessarily be accessible for another. Hence, the location that is most accessible to one firm's raw material suppliers may be very different from that which is most accessible to its customers; equally, one firm's accessiblity either to raw materials or to customers may well be different from that of any other firm. Successively higher bids are consequently made for the sites that different entrepreneurs regard as the most accessible *for their particular purposes* (section 7.3.2).

Once established, land values tend to be self-perpetuating due to competition between prospective buyers. Low-density housing, for instance, is basically more expensive than high-density housing because it requires more land to house one individual. Once such an area has been developed, there will be many people trying to buy such properties (because they are thought to be, in estate agents' jargon, 'superior'); many buyers seeking few properties means that the price goes up ... and up ... and up! Hence, low-density areas of housing *increasingly* become high land-value areas.

Evaluation inevitably entails a close analysis both of the economic and physical characteristics in order to allocate land effectively between alternative uses. Not infrequently, however, conditions of accessibility become the dominant aspect of the decision-making process: if a particular place is regarded as inaccessible (by whatever criteria), its potential use is minimised even though it may be well suited to a particular use in all other respects. The 'peripheral' areas of Britain are often excluded from consideration as land suited to population development, merely because of their remoteness (figure 4.1; Cracknell, 1967). Conversely, if a plot of land is highly accessible for any particular purpose, its physical characteristics (good or poor) may be dwarfed into insignificance and evaluation dominated by consideration of its accessibility alone.

FURTHER READING

Allen (1959); Burton (1971); Burton and Kates (1965); Clawson, Held and Stoddard (1960); Hall (1965); Mansfield (1971); Moss (1963); O'Riordan (1971B); Paterson (1972); Rothenberg (1967); Simmons, I. G. (1966); Simmons, I. G. (1973); Stewart (1968); Thomas (1973); Thomlinson (1967); United States National Research Council (1969); Waller (1970); Young (1973).

5
Labour

All forms of human activity require a supply of labour, but requirements vary considerably between different firms, industries and locations. The requirements of a large firm are different from those of a small firm in the same business, not only in terms of the total number of workers required, but also, in certain circumstances, in terms of their types and skills, largely because the bigger firm is better able than its smaller counterpart to derive scale economies through the introduction of specialisation and the division of labour. The small corner shop, for instance, may be staffed by one or two assistants, of either sex, who are required to perform all the range of operations necessary for the day-to-day running of the business; whereas the larger supermarket may be staffed by several assistants some of who are female supervisors, cashiers or cleaners while others are male shelf fillers, storemen, cleaners or managers. Indeed, as scale increases and the process of the division of labour continues, labour requirements become more precise and more exacting. The same basic principles apply to different industries: some—such as the iron and steel and manufacturing industries—require large numbers of predominantly male workers each with specialist skills in one of a wide range of trades; others—such as the textile or clothing industries—may require an equally large number of specialist females; but many of the industries that employ relatively few workers—such as wholesaling or packaging—may either require a far wider range of skills or aptitudes of each of its employees, or may even require no specialist skills at all.

Spatially, too, labour requirements vary considerably because of the tendency for human activity to agglomerate at certain locations that have at some time possessed an initial advantage with respect to other locations. That initial advantage may or may not have been originally related to the availability of a certain quantity or type of labour at a particular location, but through time, as successively more firms and industries move to that location in order to derive economies of scale, demand for a particular kind of labour becomes localised at that point. Spatial inequalities of labour demand consequently result from this mechanism of cumulative causation, but as requirements become intensified, the same process of cumulative causation leads, in the long run at least, to the creation of an appropriately skilled labour force in sufficient quantity at that same location. In other words, spatial inequalities in the supply or availability of skilled labour result from the creation of spatial inequalities in labour demand, a process that itself is multiplicative in effect and that forms an integral part of the processes of spatial agglomeration and locational specialisation.

The supply of labour and the requirements of different firms and industries are thus essentially heterogeneous in nature and display a high degree of spatial inequality (Rankin, 1973). Through time, the conditions of labour supply and demand have changed quite considerably, with the result that the locational 'pull' of labour has been correspondingly modified with respect to all other locational influences of scale, land, capital, transfer, demand and supply and other constraints and incentives. Such changes in labour requirements and availability have caused serious employment problems in many regions of the world. In just the same way that a dilemma of land use has been created by the countervailing effects of scarcity and growth, so an employment dilemma has been created by the progressive general reduction of total labour requirements in most sectors of the economy at a time when population growth rates have been such that progressively more labour has been made available. In both developed and developing areas alike, the effect of technology has been to increase the economic wealth and productivity of rapidly expanding populations, but at the same time it is clear that 'satisfactory growth of the national product is not sufficient to provide a guarantee against severe unemployment' (Galenson, 1971, p. 1). If, therefore, unemployment problems are to be minimised or even solved, locational and economic decisions must be based on the optimisation and rationalisation of the relationship between the requirements of different firms and industries and the supply of labour at different locations, variations in which are primarily related to the associated factors of demand for products, and the productivity, cost and mobility of labour.

5.1 DEMAND

The level of demand for the products of any firm, industry or sector of the economy directly affects the demand and supply conditions of labour, since changes in demand for products may necessitate changes in the output of those products, which in turn, may necessitate changes in the number of workers required to produce the new level of output. Increased demands for the products of any sector may, through this process, lead to an expansion of employment opportunities, whereas recessions in any sector may lead to the laying off of labour and consequently may cause an increase in labour supply as unemployment levels rise. Changes in the demand for the products of any one sector of a national or regional economic system inevitably have repercussions on the levels of demand and output of other related sectors. Most industries buy some of the products of other industries and indeed there is normally widespread interdependence between different firms, industries and sectors because it is by disintegration of processes that scale economies can often be achieved (section 3.1.3). A 'multiplier' effect consequently characterises changes in the output of any sector of the system and employment levels are subject to correspondingly multiplicative changes.

5.1.1 *Direct changes*

The level of employment in any sector of the economy is related to the output of that sector by a linear function of the general form

$$E_i = c + dX_i \tag{5.1}$$

where
  E_i = employment in sector i (in man hours)
  X_i = output of sector i

and c and d are constants that vary from sector to sector (Isard, 1960, p. 628). Similarly, the level of unemployment (U) in any region (j) at a given time (t) takes the form of a linear function

$$U_{jt} = A_t \times S_{jt} \times R_{jt}$$

and is related to the structure of economic activity (S), the level of demand for the products of the national economy and to the level of demand for the products of the local subregion (R) (Brechling, 1967).

The amount of labour required to produce any given level of output varies from sector to sector: in Utah in 1947, for instance, service, utility and trading activities required an estimated 88 man-years of labour for every million dollars' worth of output, whereas the non-ferrous metal industry required only 20 man-years for the same output level (Moore and Petersen, 1955). But any *additional* increments of labour that may be required consequent upon an increase in the desired level of output are partly determined by the extent to which the productivity of the existing workforce can be increased and by the extent to which alterations are made in wage rates (sections 5.2 and 5.3). Because productivity is related to the scale of operation of any activity, labour requirements are additionally related to the current scale of production in each sector: the additional labour requirements of a firm that experiences an increase in demand when production is approaching the scale at which diminishing marginal returns occur, are likely to be very different from those of a firm whose scale is such that further output would lead to the possibility of deriving increasing economies of scale. Similarly, the precise form of the relationship between employment and sector demand is partly determined by the fact that labour is a highly divisible commodity in that the amount of labour supplied can be precisely regulated by changing the hours worked by any number of employees. Seasonal or temporary changes in demand for the products of any sector of the economy may be met by the introduction of overtime or short time working and this may obviate, or at least modify, the extent to which it is necessary either to take on additional employees or to declare wholesale redundancies. In many firms and industries, overtime working often becomes a standard, though possibly unofficial, practice, simply because it is a less expensive expedient—even in the long run—than that of taking on additional workers to meet fluctuations in demand. (The need to pay the employment 'overheads' of insurance, superannuation, pension contributions or even employment taxes [such as the Selective Employment Tax that applied in Britain for some years in the late 1960s and early 1970s], is avoided by overtime working.) In the period 1961–72, overtime working in British manufacturing industries affected 32.6 per cent of all employees and involved an average rate of 8–8½ hours of overtime per week per employee (table 5.1). If additional labour is still required in order to meet changes in seasonal demand, casual labour may be employed on a full-time basis for the duration of the period of high demand: this is normal practice in the catering trade during summer in holiday resorts; in the building industry during construction at a particular site; and in agriculture during harvest time. (In

TABLE 5.1 OVERTIME WORKING: manufacturing industries, Britain 1961–1971. Source: *Department of Employment Gazette* (April, 1972).

	1961	1962	1963	1964	1965	1966	1967	1968	1969	1970	1971
Percentage of all operatives working overtime	31.9	28.8	39.4	34.0	34.9	35.5	33.0	35.3	39.3	35.3	30.7
Average number of overtime hours worked	8.0	8.0	8.0	8.5	8.5	8.5	8.5	8.5	8.5	8.5	8.5

Britain, 16 per cent of the total agricultural workforce is employed on a casual basis to meet seasonal variations in demand.) The employment of part-time casual labour may be another essential way of meeting the labour requirements of activities with demand peaking problems: the employment of cleaners in offices, factories, hotels or other residential institutions is normally on such a basis because the work of cleaning has to be done regularly and at certain times only, and the peaking of trade in retail establishments on Fridays and Saturdays is normally met by the employment of a sizeable part-time labour force. (In Britain in 1970, 26.4 per cent of the total labour force in retailing was employed on a part-time basis.) In agriculture too, part-time workers may be necessary to cope with such tasks as baling at harvest time, and in Britain such part-time labour accounted for 15 per cent of the total agricultural labour force in 1971.

Changes in demand for the products of any one sector thus lead to direct changes in the employment requirements of that sector, though the magnitude of the changes varies from sector to sector in association with the conditions of productivity, scale and divisibility which prevail in that sector (table 5.5). (Variations in these conditions are accounted for by changes in the values of the constants c and d in equation 5.1 above.)

5.1.2 *Indirect changes*

Every sector of the economy is dependent on a number of other sectors both for the supply of its raw materials and for the consumption of its products. The agricultural sector of the economy, for instance, is supplied with various commodities from the light manufacturing, heavy manufacturing, power and communications, transport, finance, construction and household sectors of the economy, but it also supplies its own end products to each of the same sectors. Similarly, the 'household' sector supplies the labour and other personal services needed by every producing sector, and, in turn, it buys the products of every producing sector for its own consumption. Such 'structural interdependencies' within the system can be described by means of a series of 'input–output' tables into which the value of the purchases of each sector from every other sector can be entered (table 5.2). (The actual number of sectors chosen for analysis depends on the nature of the economic system being analysed and also on the degree of detail required.)

A more refined and more useful way of presenting this picture of sectoral interdependencies is to standardise the data in the input–output matrix so that the interdependencies are all expressed in terms of their *relative* contribution to a standard unit of output of each industry. Consider the first column of table 5.2; the total value of the purchases of the agricultural sector is £12 210 000 and the contribution of each of the eight sectors of the

TABLE 5.2 INPUT-OUTPUT MATRIX: value of flows between sectors (£1000).

Sector producing \ Sector purchasing	Agriculture	Manufacturing	Power	Transport	Finance	Construction	Household
Agriculture	12 210	52 800	1 261	880	1 860	10 365	12 500
Manufacturing	37 851	136 628	686	1 265	678	18 930	189 540
Power	7 326	48 300	2 612	1 200	7 980	650	103 000
Transport	19 536	36 702	230	2 486	3 644	8 625	12 000
Finance	2 442	1 089	3 975	1 238	5 680	930	54 500
Construction	1 221	411	1 036	316	10 300	112	11 760
Household	41 514	44 670	14 800	41 115	62 158	17 088	34 800
Total purchases	122 100	320 600	24 600	48 500	92 300	56 700	530 200

economy to those purchases is also indicated in terms of their monetary value. If the value of each of these inputs is divided by the value of the TOTAL OUTPUT (£122 100 000), the relative contribution of each sector per £'s worth of agricultural output is revealed. Table 5.3 shows the resulting standardised data if this procedure is applied to all sectors shown in table 5.2: for each £'s worth of agricultural output, for example, the manufacturing sector is seen to contribute £0.31 worth of material inputs, the transport sector contributes £0.16 and the household sector contributes £0.34 (column 1 of table 5.3). The standardised values are commonly known as the PRODUCTION COEFFICIENTS of each sector.

Consider now the effect of an increase in consumer demand (worth, say, £1m) for agricultural products. The matrix of production coefficients (table 5.3) already shows how much additional value of demand is created in every other sector consequent upon each £1 increase in demand for its own products, so that in order to estimate the increased demand in the whole economic system consequent upon the £1m increase in direct demand for agricultural products, each of the production coefficients shown in table 5.3 must be multiplied by one million. Similar calculations could then be made for each of the other sectors: if, for example, consumer demand for manufacturing products were expected to increase by £8m, the production coefficients of that sector would each be multiplied by eight million to show the extent of additional demand created in each dependent sector. If all of these inputs are then added up for each sector (by adding the elements of the appropriate row of the matrix), the extent of the direct increases in demand for the products of each sector will be revealed (for example, had there been increases of £1m in each sector, the resulting agricultural demand would have

TABLE 5.3 PRODUCTION COEFFICIENTS (MATRIX X): direct inputs per £1 of output. Derived from table 5.2.

Sector producing \ Sector purchasing	1 Agriculture	2 Manufacturing	3 Power	4 Transport	5 Finance	6 Construction	7 Household	8 Total Inputs (excluding household sector)
Agriculture	0.100	0.160	0.051	0.018	0.020	0.183	0.235	0.532
Manufacturing	0.310	0.431	0.028	0.026	0.007	0.334	0.358	0.806
Power	0.060	0.150	0.106	0.024	0.087	0.012	0.194	0.328
Transport	0.160	0.114	0.009	0.052	0.040	0.152	0.023	
Finance	0.020	0.003	0.162	0.026	0.062	0.016	0.103	2.334
Construction	0.010	0.001	0.042	0.006	0.111	0.002	0.002	
Household	0.340	0.141	0.602	0.848	0.673	0.301	0.065	
Total	1.000	1.000	1.000	1.000	1.000	1.000	1.000	

been £0.532m). These direct requirements are known as the FIRST ROUND of input requirements. It should be noted that the household sector is excluded from these calculations because otherwise double counting would occur (Cumberland, 1960, p. 332; table 5.3, column 8).

This, however, is not the end of the effect of the initial changes in demand because the first round requirements also have to be produced and these, in turn, lead to a further round of INDIRECT DEMAND increases. Initial demand changes led to a first round increase of £0.532m for the products of the agricultural sector; in order to produce these agricultural products, a further increase in demand for the products of all other sectors is required, just as it was required to produce the initial increases, and so a SECOND ROUND of input requirements arises. This second round of increases also occasions a further set of demand increases (a third round) which in turn occasions another set of increases (a fourth round) which begets a fifth round, which begets a sixth round and so on, almost *ad infinitum*. The size of each successive round of increases soon begins to diminish however, so that after about five or six rounds, the additional demands become relatively insignificant.

In order to estimate the *total* (that is, direct and indirect) changes in the value of demand in each sector consequent upon the initial changes in consumer expenditure, it is necessary to calculate and sum the round-by-round changes outlined above. This is a lengthy procedure that becomes lengthier as more and more sectors are included in the analysis. However, a short-cut method for performing the round-by-round calculations is available and is based on the calculation of the INVERSE MATRIX (B) of the original matrix of production coefficients (X). The inverse matrix is calculated by use of the formula

$$B = (I - X)^{-1}$$

where I is the identity matrix of the original matrix X. This method obviates the need to use a power series of matrices in order to calculate the appropriate n-round coefficients, and is identical to the method described earlier for the identification of nodal structures based on n-order interdependencies (section 1.2.1).

This matrix of direct and indirect income effects is then used to calculate the total effect of any given demand changes in the economy, as Moore and Petersen (1955) have shown in their study of Utah. Suppose that there is an increase of £1m in demand for the agricultural products of the region described in tables 5.2 and 5.3: it can be seen from the matrix of production coefficients (table 5.3) that there would follow an increase of £532 000 in income originating in this industry. The indirect effects can then be assessed by multiplying each of the elements of the 'agriculture' row in the inverse

TABLE 5.4 INVERSE MATRIX B: based on table 5.3.

	Agriculture	Manufacturing	Power	All others (excluding household)
Agriculture	1.8395	0.0026	0.0074	0.0045
Manufacturing	1.2461	0.0134	0.0005	0.0623
Power	0.2241	0.0361	0.0028	0.0147
All others (4–7)	0.4211	0.1246	0.3140	0.6222

matrix B (table 5.4) by the appropriate input-output coefficient of that sector, and then summing each of the resultant figures. In this case, the calculations are

$(1.8395 \times 0.532) + (0.0026 \times 0.806) + (0.0074 \times 0.328) + (0.0045 \times 2.334)$
$= 0.99$

In other words, the effect of the initial £1m increase in demand for agricultural products is to cause a total increase of £0.99m in the income originating in that sector—a level that is £0.46m higher than the original *direct* effect alone which created only £0.53m income. This multiplier effect can be described quantitatively by expressing the *total* direct and indirect income change by the direct change; in this example the income multiplier is 1.87 (that is 0.99/0.53).

The consequences of such direct and indirect demand changes on employment can then be assessed by using the inverse matrix (B) and the relationship that is known to exist between employment and production in each sector (table 5.5): the additional employment requirements for every unit of additional demand in any one sector are estimated by multiplying each element of the appropriate row of the inverse matrix (B) by its associated *direct* employment coefficient—the values calculated from equation 5.1 and shown in table 5.5—and summing the total (an operation which is similar to that involved in the calculation of total demand changes

TABLE 5.5 EMPLOYMENT COEFFICIENTS: number of employees required per £1m of output in different sectors of an economy (hypothetical).

Sector	Employees
Agriculture	85
Manufacturing	34
Power	14
All others	386

already outlined). For example, the additional employment needs of the manufacturing sector described in table 5.4 consequent upon a £1m increase in demand for its products is calculated as follows

$(1.2461 \times 85) + (0.0134 \times 34) + (0.0005 \times 14) + (0.0623 \times 386) = 130.40$

The original increase in demand leads to the additional employment of 131 men in that sector, which is 46 more than the direct change alone would have produced, so that the direct and indirect employment multiplier is 3.85 (131/34).

The multiplier varies in its effect from sector to sector depending on the extent of the structural interdependencies that characterise each sector. Moore and Petersen (1955) thus found that the greatest multiplier effect (3.05) in the 1947 economy of Utah was experienced in the non-ferrous metals sector, and the smallest (1.20) in the utilities, trade and services sector.

Several technical and conceptual difficulties arise in trying to assess the extent of the multiplier effect (Moses and Schooler, 1960, pp. 189–205). In particular, successive increases in demand are not necessarily met by a directly proportional increase in the number of workers taken on (section 5.1.1), but in the analysis outlined above it is assumed that the relationship is homogeneous (that is, that the change is constantly proportional at all levels of demand). It is because of this, that the technique of multiplier analysis has been heavily criticised. In more refined analysis, however, attempts are made to introduce non-homogeneous employment and income coefficients that account for these variations.

5.1.3 *Induced changes*

In addition to indirect changes in demand and employment caused by the sectoral interdependencies of the system, a further series of INDUCED CHANGES may also be instigated. Suppose that an initial fall in demand for iron and steel leads to a reduction in demand for all the indirectly related industries such as machinery, metalworking, transport equipment and electrical goods. As unemployment in each of these sectors increases, more people find that they have less money available for spending generally—in other words, a change in consumer expenditure is experienced. In effect, this means that demand for certain commodities (for example, social facilities, luxury items and even food) is further reduced and this in turn leads to a further reduction in demand for the products of the various dependent industries, which ultimately induces further changes in employment requirements and leads to a further increase in the multiplier effect. Moore and Petersen (1955) calculated that the Utah non-ferrous metal industry multiplier was increased from 3.05 to 5.40, and the utilities multiplier was increased from 1.20 to 1.75 as a result of such induced changes in employment.

The total multiplier effect is thus made up of both direct, indirect and induced employment changes and varies from sector to sector according to the extent of sectoral interdependencies. A small change in demand for one product may consequently have far reaching effects, as Barfød (1938) discovered in Aarhus where the closing of an oil factory employing 1300 workers eventually led to the loss of some 10 000 jobs in a wide range of industries and services in and around that city. Sectoral interdependencies vary regionally with the result that different regional employment multipliers characterise the labour requirements of different regions. While, therefore, it may be generally true that it is those activities that have high direct employment changes per unit of demand change which also have high indirect effects (Hirsch, 1959), there may be several notable exceptions that vary regionally (Isard and Kuenne, 1953; Moore and Petersen, 1955).

5.2 Productivity

Increases in the productivity of labour have characterised most economic activities since their inception though there are marked differences between nations and regions in their rates of increase. In America during the first half of the nineteenth century, productivity per man-hour increased by 25 per cent, in the second half by about 100 per cent, and so far in this century it has increased almost threefold (Slichter, 1961). Within this broad framework the rate of increase is now beginning to decline: whereas in the years immediately after the Second World War annual increases in output per man-hour averaged 6.2 per cent, by 1965 it had been reduced to 2.8 per cent and by 1970 it was as little as 1.2 per cent. Similarly, in Britain during the 8-year period from 1963 to 1971 the average annual increase has been 4.2 per cent, whereas in the post war period from 1947 to 1955 the average increase was of the order of 8 per cent. This tendency appears to typify productivity generally in that the highest rates of productivity increase are often recorded 'in the early stages of cyclical recovery when unused human and capital resources are put to use' (Bloom and Northrup, 1969, p. 370). Productivity also varies from activity to activity: since the war, the greatest increases have normally been made in agriculture, mining and the transport and public utility industries, while the construction and retail industries have experienced relatively small increases (table 5.6). Again, however, regional variations are apparent: in Britain, productivity in the public utilities sector has been increasing at an average rate of 8.4 per cent, whereas in America the increase in the same sector has been in the order of 6.2 per cent.

Changes in productivity such as these have affected the demand and

TABLE 5.6 PRODUCTIVITY INCREASES: sectoral changes in labour productivity, Britain 1963–71; base year 1963, index number 100. Source: *Department of Employment Gazette*, (May 1972).

Sector	1961	1965	1966	1967	1968	1969	1970	1971
Agriculture	104.5	107.4	114.7	124.3	133.4	158.1	165.2	176.6
Gas, electric, water	103.5	108.8	110.0	113.8	124.1	137.2	150.1	167.6
Textiles	106.0	110.4	111.7	116.9	134.8	137.4	145.1	157.2
Engineering	106.1	106.6	112.7	117.5	124.1	128.1	130.7	137.0
Mining and quarrying	103.9	105.0	106.5	111.1	118.9	124.1	128.8	135.3
Transport	106.5	108.9	116.4	120.0	124.6	126.8	130.2	133.4
Manufacturing	107.2	109.6	111.3	114.4	122.4	125.0	126.8	131.0
Metal manufacturing	108.4	111.2	107.0	105.7	114.3	117.1	116.5	111.4
Construction	102.6	102.9	104.1	105.4	105.7	108.7	110.6	110.9
Retailing	101.7	103.4	104.7	106.8	107.0	107.9	109.4	109.9

supply conditions of the labour market by altering both the amounts and types of labour required by different activities. A major effect of the substitution of capital for labour is that the size of labour force needed to produce a given level of output is normally reduced and, indeed, this may even be the primary object of some substitution schemes especially when labour costs are high. The kind of labour needed also changes as the result of the introduction of machinery to the production process: in some circumstances the labour required to operate the machines may need to be more skilled than previously, so that higher educational levels may be essential for the maintenance of high levels of productivity; on the other hand, the widespread introduction of automation and machines may create a greater demand both for unskilled and female labour since the amount of skilled or heavy work (which previously may have necessitated male labour) is thereby reduced.

In either case, an imbalance of labour demand and supply may result at the locations experiencing productivity increases. Socially, unemployment may cause the most immediate concern, but at the same time the longer-term prospects of the region may be in jeopardy since it may not be possible to generate the newly required *kind* of labour in sufficient quantities at an old-established location. In such circumstances, the relocation of economic activity at locations where such labour is available may be necessary and the consequent problems of economic decline and stagnation and regional employment imbalance may ensue. To some extent, however, the fact that labour is essentially mobile both between industries and locations may counteract any need to relocate; in particular, the increasing possibility of the daily mobility of labour being geographically extended, may allow a greater flexibility in the entrepreneur's choice of location. In the past, when it was not feasible for workers to travel so far each day, the choice of location was rather more circumscribed since the labour supply could only be drawn from a limited area. Even so, increased commuting times do not necessarily bestow unreserved benefits to location or production. Labour productivity is lessened when commuting is increased, as Ernest Bevin suggested in 1940: 'When you can get a person from home to factory in ½–¾ hours as against 1–2 hours, you increase productivity by 9–10 per cent'. Similarly, each additional hour's travel per day increases both the days lost from fatigue and sickness—by as much as 2 per cent of earnings—and the rate of labour turnover—by about 1 per cent (Liepmann, 1944).

5.2.1 *Introduction of machinery*

Productivity increases are not primarily or necessarily solely the result of increased direct effort from each worker, though this may be a contributory factor. The efficiency of each individual worker may be increased by the introduction of specialisation and by the process of the division of labour, both of which are related to the derivation of internal scale economies in the production process (section 3.1.1), but the greatest increases in productivity result from the application of technological innovations to the productive process, and in this respect productivity is related to the substitution of capital for labour.

The introduction of machinery to the production process reduces the number of man-hours of labour required to maintain a given level of production though the extent of the reduction varies from process to process. In Thailand, for example, the introduction of tractor farming to rice production had the greatest impact on the labour inputs needed for land preparation; sowing, planting and day-to-day care were hardly affected at all, while the needs for harvesting and threshing were only slightly reduced. Rice yields (both for transplanted and broadcast rice) were also directly related to the level of mechanisation, the precise nature of the relationship being described by a series of regression equations; see table 5.7 (Inukai, 1971).

TABLE 5.7 RICE YIELDS AND MECHANISATION: regression equations describing the relationship between mechanisation x and yields of transplanted rice Y_t and broadcast rice Y_b; Central Region of Thailand. Source: Inukai (1971).

Farm size (rais)	Regression Equation	Correlation (r^2)
15–30	$Y_t = 219.12 + 1.24x$	0.45
	$Y_b = 153.51 + 1.33x$	0.59
31–45	$Y_t = 206.93 + 1.14x$	0.87
	$Y_b = 154.94 + 1.00x$	0.45
46–60	$Y_t = 193.28 + 1.01x$	0.59
	$Y_b = 146.40 + 0.98x$	0.42
61–140	$Y_t = 182.85 + 1.06x$	0.60
	$Y_b = 117.85 + 1.07x$	0.53

Similar effects of the introduction of innovations have been felt in a wide range of industrial activities. Osborn (1953) has shown that the introduction of automation to a part or to the whole of a process leads to a reduction not only of the number of employees required but, importantly, to the labour costs per unit of output. Twelve American manufacturing and non-manufacturing activities were analysed ranging from a continuous-flow chemical process, several food processing operations, to certain clerical and administrative functions. There were many variations in the extent to which automation caused a re-appraisal of labour needs; the effect of introducing a

99 per cent automatic process for the production of napalm in one of the Ohio plants of the Ferro-Chemical Corporation in 1951 was to reduce the employee requirement by 77 per cent; whereas the automation of only a *part* of the lard-rendering processes by an Indianapolis firm led to a reduction of only 50 per cent of the labour force. There is also great scope for capital/labour substitution in clerical and administrative operations, since the introduction of computers makes possible the shorter completion of some clerical tasks that might otherwise require very large amounts of labour for very long periods of time.

5.2.2 *Education and social services*

Productivity is undoubtedly also related to the continuing process of education, and indeed economic growth cannot proceed without increases in the general educational level of the labour force. It is for this reason that education is regarded as a primary objective in all development plans. Many attempts have been made to show the general principles of the relationship between the requirements in educational standards of the workforce and economic and productivity growth rates. Tinbergen (1963), for instance, has shown how different educational requirements at both secondary and tertiary (higher) levels are entailed in producing different rates of economic growth and labour productivity. Of necessity, such a model has to make certain working assumptions about teacher–student ratios, the study period involved at each educational level, and the output per worker with given educational qualifications; and these assumptions vary according to the country or area being analysed. In the USA a national output growth rate of 4.47 per cent per annum (30 per cent per 6-year period) would require a 118 per cent increase in the number of students completing a secondary-school course and a similar increase in the number of students completing higher level courses. The greater the required economic growth or productivity increase, the greater is the required educational background of the workforce—a growth rate of 5.77 per cent per annum in national output thus calls for a 175 per cent increase in secondary-school students and a 180 per cent increase in tertiary level students.

Layard and Saigal (1966) have analysed the correlation between differences in labour productivity in thirteen countries and the distribution of the labour force within sectors, the educational attainments of the labour force and the proportions of certain high level minor occupational groups in the national labour force as a whole. A 1 per cent increase in productivity in the economy as a whole appears to be correlated with certain percentage increases in the proportions of workers who have completed secondary school, attained matriculation level or completed only primary schooling; and also with a 0.81 per cent increase in the proportion of architects, engineers, surveyors, scientists, draughtsmen and technicians in the national labour force. Productivity increases in different sectors are associated with different educational changes: a 1 per cent increase in the productivity of the manufacturing and electricity sector, for instance, is associated with a 1.02 per cent increase in the proportion of the labour force having completed secondary or higher education courses and a 0.76 per cent increase in the proportion of those having had middle schooling.

The effects of such educational measures as raising the school-leaving age, providing vocational training in schools, expanding the opportunities for further education generally and providing in-service training schemes or retraining schemes, are all inevitably reflected in the quality of labour and also, it should be noted, in the cost of labour. As the educational level of an individual increases, so his average earnings expectancy also increases and is reflected in the so called 'age–earnings' profiles of different individuals. Although these profiles differ in detail from country to country, the general principle that increased wages and salaries over time are correlated with increased educational level appears to apply in most regions (figure 5.1; Blaug, Preston and Ziderman, 1967).

Most countries have now established sizeable education programmes and there have been marked increases in education provision that are ultimately reflected in the quality of the employed workforce. Within the period 1952–66, for example, the proportion of American men employed who had

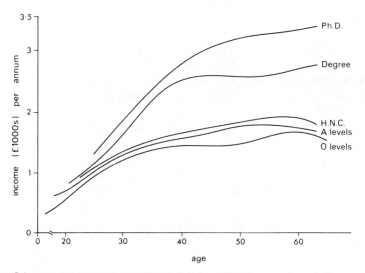

Figure 5.1 AGE–EARNINGS PROFILES Britain, 1965. Source: Layard, Sargan, Ager and Jones (1971).

completed 4 or more years at high school had increased from 40 per cent to 57 per cent. Because such educational programmes appear to result in increases in productivity, it would seem that 'those countries with high investments in post-primary education ... are laying firm foundations for economic growth and cultural change, whereas countries with low investments in this field will continue to be confronted with grave problems in the mobilisation and utilisation of their human resources' (Ginsburg, 1961, p. 44). The provision of education at different levels is, however, a costly business. Quite apart from the tangible capital costs of educational buildings and their maintenance, the recurrent costs per pupil increase as their study programme continues; in the 1960s in Britain, for example, the costs increased from £76 for a student who left secondary school at the end of the year after his fifteenth birthday; to £385 for a student who left school at 15 but followed a part time course up to the age of 18; to £2502 for a student who stayed in full time education through to graduation from a university at the age of 21 (Blaug, Preston and Ziderman, 1967, p. 93).

The provision of housing, health and welfare services may, in association with education programmes, also lead to the establishment of a labour force which, by being healthy and well cared for, can contribute effectively to the process of growth through the higher levels of productivity that it can then maintain. Absence from work due to sickness, even in regions where there are good housing, health and welfare services, may cause a sizeable loss of man-hours and thereby cause a reduction in general productivity. In Britain, accidents, psychosis, influenza, bronchitis and nose or throat infections are the most common causes of incapacity for work and lead on average to a loss of 7.8 days absence from work per year for each male worker under the age of 45; females tend to be absent rather more frequently (an average of 10.3 days), one of the more common causes of absence being pregnancy. Absence tends to increase with age, but in most age–sex groups the sickness rate tends to be lowest in East Anglia and southeast England and increases northwards and westwards across the country (figure 5.2).

Social services are expensive to provide and maintain and it is for this reason that they are rarely supplied extensively by individuals or groups in the private sector. Some firms, it is true, do provide some housing, health and welfare services but the basic responsibility for the provision of these services ultimately must remain with the state. In Britain, housing subsidies amount to some £18 500m annually (section 9.2.3), while the National Health Service, which basically exists to provide adequate medical care and attention for those who could not or do not wish to afford private treatment, costs just over £1500m each year (approximately 4 per cent of the Gross National Product).

As well as requiring large amounts of capital in any one year, all these

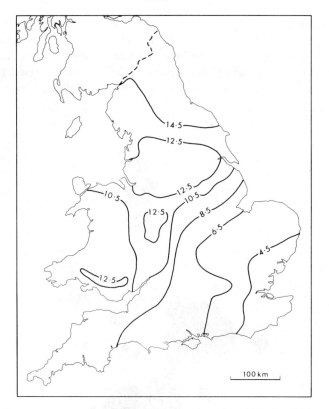

Figure 5.2 VARIATIONS IN ABSENTEEISM number of working days lost per worker (Britain, 1971).

schemes will only be productive over a relatively long period of time. The return on capital is thus not sufficiently direct or quick enough to attract large proportions of private investment funds and, in that event, housing, education, health and welfare have, of necessity, to be supported from public funds (section 6.2.2).

5.2.3 *Industrial relations*

Productivity is undoubtedly related to the state of industrial relations that may prevail in different firms, industries and regions. Militancy and industrial stoppages are often more pronounced in older industrial agglomerations than in small towns or rural and 'fringe' areas, and this may help to make the latter areas appear more attractive to certain employers. In Britain in 1971, prominent stoppages due to industrial disputes caused the loss of some 13 million working days, of which the majority were located in the established conurbations (figure 5.3). Naturally, it is not surprising that there should be such a high correlation between disputes and established industrial areas, since the largest employers are obviously located in such areas, but the greater sense of community and the development of much closer contacts between management and labour that tends to characterise the smaller firms may well

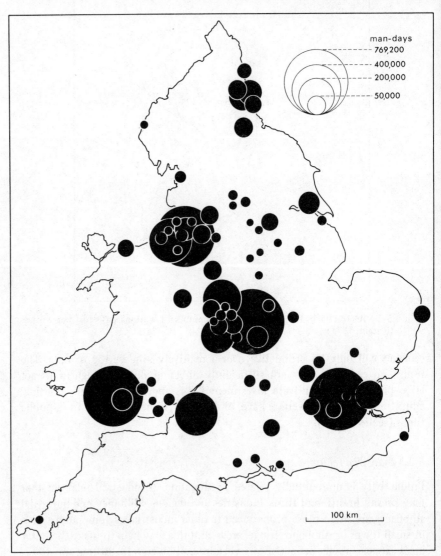

Figure 5.3 INDUSTRIAL STOPPAGES number of man-days lost through industrial disputes (England and Wales, 1971). Source: *Department of Employment Gazette* (April, 1972).

be one of the reasons why stoppages are apparently less frequent in newer or more remote industrial areas. The development of 'solidaristic orientation' (Goldthorpe et al., 1968), in which work is seen as a group activity, leads to better industrial relations, fewer stoppages and, consequently, to increased productivity. For this reason, governments may attempt to introduce various forms of legislation designed to improve industrial relations generally (section 9.3.1).

5.2.4 *Substitution*

Increases in labour productivity may thus necessitate increases in capital investment not only in machinery but also in the provision of educational and other services. In many systems such increases in capital investment may not be feasible because of the lack of funds (chapter 6). Furthermore, the possibility of factor substitution between capital and labour is dependent on the relative costs of the two inputs with respect to the marginal product that they may create (section 3.1.2). In situations where labour is expensive or scarce and capital is relatively inexpensive and plentiful, the introduction of machinery is obviously more likely to be an economic proposition than in situations where the conditions are reversed, and it is for this reason that it is often argued that the 'direct transfer of technologies from industrialised to developing countries may be inappropriate' (Marsden, 1971, p. 115). Certainly the diffusion of innovations is closely related to the decision maker's expectation of their profitability which in turn is conditioned by the ratio of capital to labour costs. In the American Midwest, for instance, the introduction of mechanical reapers to the harvesting process was accelerated by a dramatic increase in labour costs that occurred during the period 1853–54 as a result of increased demands for labour while labour was in relatively short supply. Prior to this increase in labour cost, only very large farms were able to substitute the reaper for labour, because below a certain size 'threshold' the total cost of a reaper exceeded the potential reduction in wage costs that its introduction would allow: the purchase of a reaper (the McCormick hand-rake reaper) in the period 1849–53 was equivalent to the hire of 97.6 man-days of labour, which meant that it was an uneconomic proposition for farms of less than 46.5 acres. The result of the increase in labour costs was such that the purchase of a reaper had fallen to an equivalent of 73.8 man-days of labour in the period 1854–57, and consequently more farms were able to substitute machine for man (the threshold size was lowered to about 35 acres) (David, 1966).

Labour productivity therefore relates to the opportunities of substitution that are directly related to the available level of technology in the system, and at any given level of technology there exists a relationship between output and labour inputs that changes as technological innovation proceeds.

TABLE 5.8 PRODUCTION COSTS: manufacturing industries and agriculture, Britain. Sources: Census of Production (1968), H.M.S.O. (1971) and *Annual Farming Review* (1971) (Cmnd 3229).

Industry	Total output value (£1000)	Wages and salaries value (£1000)	Wages and salaries % of output	Materials and Fuels value (£1000)	Materials and Fuels % of output	Transport value (£1000)	Transport % of output	Capital Investment value (£1000)	Capital Investment % of output
Coal mining	733 598	365 640	49.8	170 917	23.3	16 822	2.3	53 941	7.4
Quarrying	128 660	23 242	17.9	22 159	17.1	21 122	16.3	16 521	12.7
Grain milling	310 521	23 598	7.6	205 334	66.1	7 243	2.3	8 373	2.7
Milk production	258 465	17 895	7.0	138 171	54.5	7 515	3.0	4 855	1.9
Fruit and vegetables	264 308	38 459	14.6	148 398	56.1	8 554	3.2	11 079	4.2
Brewing	724 340	67 334	9.3	248 143	34.3	5 877	0.9	32 215	4.4
Mineral oil refining	506 334	21 342	4.2	439 950	86.8	844	0.2	11 913	2.3
Iron and steel	1 317 012	228 795	17.4	867 739	65.9	40 457	3.1	101 636	7.7
Man-made fibres	209 854	32 575	15.5	97 240	46.3	4 032	1.9	11 730	5.6
Shipbuilding	424 862	157 404	37.0	165 122	38.9	42 040	9.9	16 043	3.8
Lingerie and dresses	216 053	52 768	24.4	110 958	51.4	15 977	7.4	4 049	1.9
Gas	480 556	101 156	21.0	211 750	44.1	10 422	2.2	89 345	18.6
Electricity	1 696 311	276 585	16.3	528 838	31.2	827	—	554 765	32.7
Water	179 423	48 875	27.2	37 958	21.2	163	—	67 264	37.5
Agriculture	1 997 500	305 000	15.3	749 000	37.5	225 000	11.3	245 000	12.3

5.3 COST

The cost of labour is measured in terms of wage rates, and although a minimum level of labour is necessary for the production of any commodity, the average employer is concerned to keep wage rates, iike transfer, land and capital rates, at as low a level as possible so that his product can be marketed at as low a price as possible and his profits can be maximised. For the vast majority of industries the cost of labour forms the second largest item in the structure of production costs, and it is rare that labour costs are dwarfed into relative insignificance by the costs of raw materials, capital overheads or transfer (tables 5.8 and 5.9; McClelland, 1966; Chisholm, 1966; Estall and Buchanan, 1966). Consequently, most firms and industries are particularly careful to economise on labour wherever possible, and pay close attention to the wage rates that they offer.

TABLE 5.9 LABOUR COSTS IN RETAILING. Britain. Source: Report on the census of distribution and other services (1966) H.M.S.O. (1970).

Type of Store	Sales Turnover (£1000s)	Total Number of Employees	Wages and Salaries (£1000s)	(as % of turnover)
All retailing	11 638 834	2 811 917	1 185 462	10.14
Grocers	2 399 306	463 002	132 578	5.53
Butchers	665 871	130 050	56 111	8.42
Dairymen	315 212	54 061	41 771	13.25
Confectioners, tobacconists and newsagents	1 023 942	298 181	48 153	4.70
Clothing and footware	1 706 379	431 163	193 039	11.31
Household goods	1 324 013	333 810	184 053	13.90
Chemists	490 895	120 877	59 687	12.16
General stores	1 298 254	326 032	162 798	12.54
Department stores	548 174	161 291	89 010	16.24
Mail-order houses	383 664	45 667	23 118	6.03

The rates that are offered depend on the conditions of labour demand and supply that characterise different sectors of the economy at different locations and at different times. Where demand exceeds supply, wage rates may have to be increased in order to attract an adequate supply of labour, though the extent to which this is possible may be determined by the extent to which labour is 'mobile' either between different jobs or between different industries. In general, the supply of labour between locations is far less 'elastic' (section 8.1) than the supply between firms and industries, simply because labour is, in the short run, less mobile geographically than it is

TABLE 5.10 INTER-INDUSTRY AND INTER-REGIONAL WAGE RATES: average weekly earnings of manual workers (£) (3rd week pay), April 1969, Britain. Source: Department of Employment (1971), *British Labour Statistical Yearbook*, table 14.

Industry	Region							
	SE	E Anglia	SW	W Midland	E Midland	Yorkshire	NW	N
Food, drink, tobacco	24.74	21.17	22.53	24.17	21.78	21.57	23.45	21.66
Chemicals	24.79	23.31	23.84	24.23	22.02	22.60	25.99	25.42
Engineering	24.30	22.20	22.87	25.57	23.81	22.63	23.24	24.30
Shipbuilding	26.65	20.77	24.87	—	—	25.40	26.04	25.09
Vehicles	28.75	23.36	25.43	31.61	24.93	24.26	25.83	27.00
Paper, printing, publishing	30.51	24.94	25.01	25.86	24.03	23.83	26.79	25.83
Mining and quarrying	24.81	—	22.03	25.44	24.92	23.82	25.16	23.51
Transport	36.73	23.99	22.61	25.38	23.68	23.63	24.72	22.08
Public Administration	19.73	16.68	17.61	19.14	18.28	17.38	18.30	17.65

between firms or industries. Inter-firm and inter-industry wage-rate differentials thus tend to be greater than inter-location differentials (table 5.10); even so, wage rates are normally highest in the largest towns and cities of any region because of the increased competition existing between firms and industries for the limited supply of labour available at those locations; but high rates may also be offered in remote areas in order to try and attract labour to them.

5.3.1 *Demand*

There is little reason to suppose that labour is any different from any other factor of production in that demand varies in direct proportion to its price, but it is notoriously difficult to chart the precise form of this relationship.

It is apparent that employers feel that they can profitably employ more labour at low wage levels than they can at high wage levels, but there are limits both to the maximum and minimum amounts of labour that can physically be employed. The existence of scale economies permits the useful employment of extra units of labour up to a certain point beyond which diseconomies in the form of diminishing marginal returns begin to set in (section 3.1.1); it follows that beyond this critical point, additional units of labour will not be employed, no matter how low the wage rate falls. Long before that point is reached, demand for labour begins relatively to decrease as prices are lowered, simply because of decreasing marginal returns in terms of output to labour input. Conversely, a minimum amount of labour is necessary for any form of production: the number of cleaners required to provide a basic minimum standard of cleanliness, for instance, cannot be reduced beyond a certain level without a complete revision of the standards required. Any change in wage rates beyond this 'threshold' level simply has to be met by the employer and in such a situation it is clear that the demand for labour is essentially inelastic. The demand curve for labour thus takes a generally downward sloping form, though the threshold level is characterised by a horizontal line and the curve generally becomes progressively more horizontal towards the right (figure 5.4).

Changes in the economic or technological state of the nation, sector, industry or firm, together with changes in labour productivity, may cause entrepreneurs to employ either more or less labour at different wage rates. Shifts in the labour demand curve thus result: an outward shift in the curve (represented by the curve $D'D'$ in figure 5.4) means that *at every wage rate* the employer is willing to hire more labour than previously, whereas a downward shift (represented by the curve $D''D''$ in figure 5.4) means that *at every price* the employer is willing to employ less labour than previously.

In order to acquire an adequate supply of labour for any level of

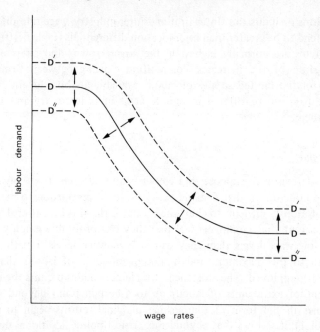

Figure 5.4 LABOUR DEMAND the relationship between wage rates and the demand for labour.

production it may be necessary for any individual firm or industry to vary the wage rate that it can offer in relation to the rates offered by other competing firms or industries who may also be seeking to satisfy their requirements from the same source of supply. If the total available supply of labour is limited for any reason, the firm may find it necessary to increase its wage rates in the hope of attracting labour from other firms or industries and possibly from other locations.

5.3.2 *Supply conditions*

The level of unemployment at any given time in any particular area is an important element of the supply conditions of labour: when unemployment is at a very low level, wage rate variations may well be an essential mechanism by which different firms are able to attract sufficient labour for their levels of output. Normally, it is the larger firms that will be most affected, since they may need to attract a greater percentage share of the available labour force. The small firm may well be able to satisfy its relatively small needs without changing its wage rates even in times of full employment because it may be able to attract its employees for reasons other than financial ones; for example, it may be regarded as a 'good' employer, or its working conditions may be better.

On the other hand, when labour is plentiful, employers may be able to meet all their labour requirements without having to vary the wage rates offered, since there will be no direct competition from other firms or industries. In such a situation, wage rate differentials between firms will be established according to such criteria as conditions or difficulty of work, and the supply of labour consequently becomes more 'elastic'.

Workers may, however, be rather reluctant to move from one firm to another for merely financial reasons: they may, for instance, be quite happy with their present employer, they may feel that the new job would involve increased travel to work or a lot of unnecessary upheaval in finding a new home, or they may feel that they might lose status or seniority with their present employer. Equally, if the move involved a slightly different job, retraining may be necessary, and the individual may feel that even the financial incentive to move was just not worth it (see section 5.4). A very sizeable increase in wage rates by the firm seeking to attract the employees of such firms might consequently have to be made if the bid is to be successful. Furthermore, the increase in wage rate that may be necessary to attract additional labour will normally have to be applied to the existing workforce of the firm (for the obvious reason that equal payments must be made for similar work completed), so that the final cost to the firm of attracting additional labour will be considerably greater than the actual costs paid to the additional workers employed (that is, the marginal cost of labour will exceed the supply price of labour if wage rates have to be increased in order to attract further employees). The effect of varying the wage rate in this way may, however, be rather short-lived since, if one firm increases its rates of pay, either the normal process of competition or the equally normal process of trade union intervention will tend to cause the other firms to follow suit in order to keep their existing workforce. Nevertheless, wage rate variations may be necessary where the required type of labour is in short supply.

5.4 MOBILITY

Most location theory is based on the assumption that labour is essentially a mobile resource: the 'cumulative causation' hypothesis, for example, is based on the proposition that once a given activity is set up at any given place it will tend to act as a locational magnet attracting not only other activities to a proximal location, but also attracting workers from other locations and from other firms or industries already in existence at that location (section 3.2.2). In fact, there is ample empirical evidence to suggest that the extent of labour mobility is, in most economies of the developed world, quite considerable

(Lipset and Bendix, 1964), but that inter-firm or inter-industry mobility is rather greater than inter-location mobility—a fact that is also reflected in the difference between inter-firm and inter-location wage rate differentials (table 5.10). In England and Wales during the 1960s only a half of the average 10 million job changes recorded each year involved the individuals concerned in moving their home from one location to another. In West Germany, during the same period, only 6.2 per cent of the employed population moved from one location to another in association with job changing, whereas 37 per cent of the population changed jobs without changing their place of residence.

One reason why geographical mobility appears *in the short run* to be less than inter-industry mobility, is that labour mobility on a daily basis has been made increasingly feasible over increasingly greater distances as transport technology has progressed over the last half century or so. In England and Wales, for example, 36 per cent of the employed population makes a daily journey to work that takes them from one local authority area to another (Lawton, 1968). The possibility of increased daily mobility may act as a partial brake on migrational moves since, if workers are prepared to spend more time and money on a daily journey to work, the need to make a limited geographical move of home in association with a change of employment will be lessened, or even obviated altogether.

Nevertheless, during this century, most countries have experienced a pattern of rural depopulation as more and more workers and their families have migrated to developing urban areas in search of better opportunities and, more recently, several countries have seen how the trend can be reversed by the introduction of deliberate policies of industrial decentralisation. To some extent, therefore, labour does appear to move in sympathy with changes in the location of job opportunities, *in the long run.*

The extent of both short-term and long-term mobility is, however, dependent on the extent to which individuals are prepared (or able) to change either job or location. The attractiveness of increased income has to be set against a wide range of other considerations that may be just as important as money in encouraging or dissuading individuals to move; indeed, it is significant that only 3.1 per cent of job changers in Britain consider remuneration in itself to be the most important factor in their motivation (table 5.12).

5.4.1 *Information*

The decision to change jobs or to migrate in search of better job opportunities is based on the individual's perception of all the available opportunities, and on his assessment of which alternative is likely to bring the greatest benefit either financially or in terms of some other scaling criteria. As in most decision making, it is rare that the individual has either complete or accurate information about the alternatives, and this is a major factor in determining the extent and the patterns of both occupational and geographical mobility.

In certain instances, imperfect knowledge may lead the would-be job changer to decide that his *status quo* is as good as anything else that appears to exist, and may consequently dissuade him from making any move. Yet in other instances, imperfect information may cause the individual to make moves that do not offer any substantial improvement in his financial or other conditions and, ultimately, he may be further encouraged to move again: in this way, the frequency of job and home changing may be increased through imperfect perception.

The circulation of information, and the contact-frequencies of individuals are both limited by distance (section 2.1.1), so that the greatest amount and the most accurate kind of information that the would-be changer is likely to possess, normally relates to opportunities that arise within a fairly circumscribed geographical area. Correspondingly less and less information can be found about opportunities at a distance, and even if it were found, it would probably be regarded as less reliable by the decision maker. For these reasons, job changes involving migration and house changing are characterised by a marked distance–decay function (figure 5.5).

Since the extent and reliability of an individual's information field is correlated with his age, education and income level (chapter 2), it might seem reasonable to expect that these same variables should be correlated with the mobility of population generally. Certainly, there is ample evidence to show that labour turnover rates are correlated inversely with age: in Britain, the New Earnings Survey of 1970 showed that while 16.2 per cent of all male employees and 25.3 per cent of all female employees had been with their present employers for under 12 months, the rates were highest for both males and females under the age of 18 (57.9 per cent and 60.3 per cent respectively), and progressively decreased so that only 4.5 per cent of males and 9.5 per cent of females between the ages of 60 and 64 had been in their present employment for less than 12 months (table 5.11). In terms of migration also, the age of the migrant is possibly negatively related to the length of migration, in that younger workers are likely to move greater distances than older workers. Olsson (1965), however, in a study of migration in Sweden, was unable to verify this hypothesis with any certainty. (Although the regression coefficients took on the expected minus sign, they were not statistically significant.)

The effect of education on mobility is rather more complex and there are marked differences between occupational and geographical mobility in this respect. Occupational mobility is certainly higher between unskilled jobs than

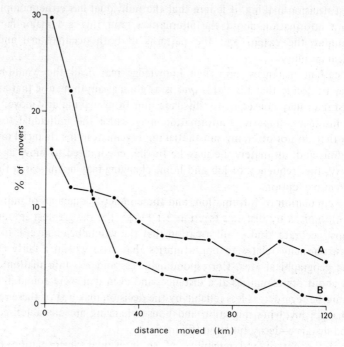

Figure 5.5 MIGRATION LAPSE RATES A—high income groups; B—low income groups (migrants to Lille, France, 1965).

TABLE 5.11 LABOUR TURNOVER: percentages of employees who had been with their employers for under twelve months, Britain. Source: New Earnings Survey, April 1970, *Department of Employment Gazette,* April 1972.

Age group	Males	Females
<18	57.9	60.3
18–20	30.3	34.7
21–24	29.9	32.4
25–29	22.9	32.7
30–39	15.8	27.5
40–49	10.6	18.9
50–59	6.4	12.2
60–64	4.5	9.5
>65	13.1	7.8
All ages	16.2	25.3

between professional occupations, and both semi-skilled and skilled jobs display turnover rates that are intermediate between these two extremes. The New Earnings Survey of 1970 in Britain, for instance, revealed that 26.0 per cent of all unskilled males had been with their employers for less than 12 months, whereas the proportions of semi-skilled male workers who had been so employed was 16.1 per cent, and those of skilled and professional workers were respectively 13.0 per cent and 6.6 per cent. By implication, therefore, the effect of education is to reduce job changing and labour turnover. On the other hand, geographical mobility tends to increase as the migrant's educational background increases: workers with 13 years background of full time education (up to the age of 18) are more liable to make migrational moves than workers with only 10 years background (up to the age of 15). The trend is, however, reversed for workers who have received any form of higher education: a relatively smaller proportion of college and university graduates make migrational moves than migrants of shorter educational background yet, paradoxically, such moves as are made by these groups tend to involve greater geographical distances than the moves of any other group (figure 5.6).

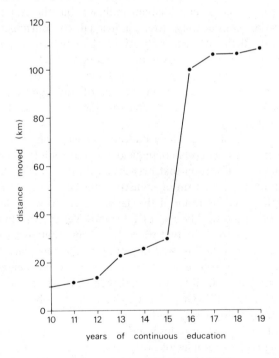

Figure 5.6 MOBILITY AND EDUCATION.

Since educational background is closely related to the income level of the individual's family, it is not surprising to find that the same general correlations between both occupational and geographical mobility and the migrant's income level are apparent as between mobility and educational background. Yet the correlations that do exist are of the most tenuous kind, and it is rare that very high degrees of statistical correlation have been found in detailed studies of labour mobility. Olsson (1965), for example, was only able to conclude that the length of a migration is *probably* positively related to the migrant's family income. That migrations tend to be of longer distances for individuals with high incomes and educational background, is at least partly due to the fact that such individuals' information fields are generally both more extensive and reliable than those of migrants with lesser incomes and education. The distance–decay function of geographical mobility consequently varies according to income and social group (figure 5.5).

5.4.2 *Evaluation*

The possibility of better prospects, whether economic or social in nature, is the motive force that generates labour mobility, but the evaluation of such prospects varies considerably from individual to individual, mainly in association with their different motivations. A number of evaluative criteria combine to define the 'imageability' (section 2.2.1) of the available prospects. Considerations of the expected income to be gained from any move must be balanced by an evaluation of any additional or indirect expenses that may be involved and of the many social and 'status' consequences that may arise in making the move.

The trade-off within the family budget between changes in gross income and changes in transport costs consequent upon a possible job change is the main economic consideration that may affect the outcome of the decision to move. To move from one firm or industry almost certainly involves a change in both the time and the route of the daily journey to work, both of which entail a change in the weekly cost of transport. Yet income and commuting cost variations can work in two ways. In certain circumstances, increased commuting costs may be involved in getting to the new place of work, but the additional wages gained more than adequately compensate for these additional outgoings so that the worker is, in net income terms, still economically 'better off'. On the other hand, some jobs that offer *less* in wages than the employee's present job, may still become attractive to the employee if the job is nearer to his home and consequently allows a *reduction* of commuting costs. Examples of both these situations were discovered in a study of job changing consequent upon the establishment of a new factory

(K Shoes) in the Furness area of Cumbria: a third of the men who lived in the village of Askam took a job at the new factory even though their take-home pay was less than in their previous job, because, among other factors, the costs of commuting were considerably reduced (Grime and Starkie, 1968). The actual amount of money spent on commuting may not, however, be the only consideration, and the individual's valuation of lost *time* may be just as significant (the fact that an average of 20 minutes could be saved on every single journey to work was a further reason why the Askam workers were persuaded to accept the lower rate of pay at the new shoe factory). Paradoxically, it appears that those who place a high valuation on time lost in commuting are precisely those who live furthest from their place of work and have the longest commuting journeys of all. It may simply be that, having lost so much time in transit, such people are more acutely aware of the problem than those who spend relatively little time in this way but, even so, it is curious that such people still continue to make the lengthiest daily journeys to work.

Similarly, the relative costs of geographical moves between the alternative locations in which prospects are thought to be good, may well affect the final decision to move from one place or other. Thus, although individuals may be prepared to move over greater distances in order to find a job in places where the overall economic prosperity is thought to be high, it seems that they will normally move to the *nearest* of those places so as to minimise their removal costs. This general principle was found by Olsson (1965) to characterise migration behaviour in Sweden: the distance over which individuals migrated was generally correlated with the size of the place of in-migration—a relationship that is not unexpected since migrants frequently imagine job prospects to be higher or better in large places than in small ones—but the potential migrant usually moves to the nearest of the alternative large places. The principle of 'least effort' (Zipf, 1949) thus appears to cause movement-minimisation in migrational behaviour in much the same way that it affects the spatial behaviour of retail consumers (section 8.2.3). Olsson found that 72.6 per cent of all migrants to larger places than their place of origin conformed with this hypothesis, and that the behaviour became more marked the larger the place of in-migration. In other words, 'it seems possible that the individual behaviour that generates the central place systems also generates the spatial distribution of migration' (Olsson, 1965, p. 38; Willis, 1972).

Many non-financial aspects of changing jobs or residence may be just as significant in determining the outcome of the decision-making process. For example, in changing from one job to another or in changing from one firm to another, the worker may well consider any possibilities of associated changes in status or security to be more significant than any marginal increase in income that may result from the move. Similarly, if the job change involves retraining for a different skill the individual may consider the extra effort or upheaval not to be worthwhile: on the other hand, the possibility of training for a skilled job may act as an additional incentive for certain workers to move. Generally, it is the younger unskilled employee who is most likely to be attracted by this possibility. It is not without significance, for example, that most middle-aged skilled workers have started work either as labourers or in only semi-skilled trades and have made two or three job changes in the process. In a study of movement into and out of skilled manual occupations in Dagenham and Battersea, Jeffreys (1954) discovered that only 40 per cent of the skilled men had entered a skilled trade on leaving school and remained in it without a break. The majority had had at least one other occupation as well as their present skilled manual trade. Some of them had started work in a skilled trade, had subsequently done other work and then returned to a skilled trade, but in most cases those with two or more previous occupations had started work in semi-skilled or labouring jobs. The largest proportion had started in juvenile jobs as errand boys, messengers or tea boys.

Job satisfaction (or dissatisfaction) may be a further and allied factor that the individual considers in his evaluation of alternative offers of employment. In all industries in Britain, about 10 per cent of employees changing jobs give 'job dissatisfaction' as the major reason for their moving (table 5.12), and in

TABLE 5.12 REASONS FOR LEAVING EMPLOYMENT: Britain 1957. Source: *Stores and Shops* (1957).

Reason	% of all leavers		
	Men	Women	Both sexes
Dissatisfaction with job	9.6	11.0	10.3
Physical working conditions	1.8	1.9	1.5
Personal betterment	21.1	12.4	16.7
Remuneration	4.2	2.1	3.1
Security	18.4	3.6	11.0
All other reasons	44.9	69.6	57.4

several instances jobs that are regarded as 'satisfying' are accepted, despite the fact that higher wages can be found in other jobs that are regarded as less satisfying. The physical working conditions that obtain in different jobs, firms or industries are also considered by some workers to be crucial considerations in their evaluation of alternative job opportunities (table 5.13) with the result that the actual wages offered may be a relatively insignificant factor in the decision-making process. Within each industry, the various conditions of work are evaluated in different ways: in the retailing industry, for example, it would appear that men consider the chance of promotion, the

TABLE 5.13 STAFF ATTITUDES: conditions of work which are most preferred – sample of employees in light engineering, Exeter, 1971.

Facilities	Preference Ranking	
	Men	Women
Good promotion prospects	1	9
Good working conditions	4	2
Good club facilities	6	5
Have pleasant colleagues	5	1
Good security	8	8
Interesting job	3	3
Good pay	2	4
Pay increases proportional to merit	7	10
Work for a well-known firm	10	7
Training facilities	9	6

possibility of improving their skills and of receiving pay increases in proportion to merit, as the three most important conditions of employment, whereas women consider it essential to have pleasant working companions, reasonable chances of promotion and good working conditions and other amenities (McClelland, 1966).

Changing residence also involves a number of non-financial considerations. In addition to the short-term upheaval of the move itself, long-term changes involving friendships, the education of children and many other aspects of day-to-day domestic life may be involved. In certain circumstances such considerations may well act as incentives to move: the lack of friends, unfriendliness on the part of neighbours, poor schools, poor social facilities and even a poor environment, generally would all act as 'push' factors pushing in the same direction as the 'pull' factor of better wages or social conditions elsewhere, and indeed many local and regional migrations have been occasioned by the thought that 'The grass is greener on the other side of the fence.' Conversely, the existence of a well-established circle of friends in a friendly and pleasant location with good schools and other social facilities may counter the attractiveness of better prospects elsewhere, particularly if the social and environmental conditions of the 'new' residence are known to be worse than those already enjoyed.

Job-changing and migration thus reflect not only the opportunities that may be available at different times and at different locations but also the personal characteristics, attitudes and information levels of the individual worker. It would consequently seem that occupational and geographical labour mobility represent the final response to a number of stimuli and that variations in wage rates are less significant in creating labour mobility than other conditions of a less financial nature that together operate as a kind of process–response system (Lövgren, 1956).

FURTHER READING

Britton (1967); Gordon (1973); Greytak (1972); Haggett (1971); Hansen (1970); Hewings (1971); Humphrys (1965); Jones (1965); Kerry Smith and Patton (1971); King, Casetti and Jeffrey (1969); Lester (1964); Miernyk (1955); Morrison (1973); Rankin (1973); Round (1972); Simmons (1968); Westergaard (1957).

6
Capital

Growth in any economic system is only made possible through the substitution of capital for the other factors of production, land and labour. Substantial increases in land and labour productivity lead ultimately to greater prosperity and to higher personal disposable incomes which in turn generate further demands for greater growth. The mechanisms by which capital is generated and allocated within the system constitute a crucial aspect of spatial organisation since it is through their operation that the intricate relationship between scarcity and growth is created, maintained or modified (section 4.1.1).

The sources of capital are the voluntary or forced savings of individuals, firms, industries or governments, and it is by investing these savings in different ways within the system that the process of growth is established. Individuals, families and households normally invest their savings in bonds, stocks, shares and securities, though they may also invest less directly by putting their monetary savings into various accounts with banks, building societies and other financial institutions. Their reasons for doing this vary considerably: some may simply wish to hoard money, others may be desperately keen to make money by investing it, while others may simply wish to save for the proverbial rainy day (whether that be old age, daughter's wedding, children's schooling or whatever). Whatever the reason for and whatever the form of the saving, the money involved is made available, temporarily, and for various periods of time, to other groups of people who may use it in a variety of ways. Firms and industries may be the beneficiaries, but they too are able to save by allocating a part of their profits to capital development projects which may involve either buying new equipment, tools, machinery and any other of their man-made factors of production, or the building of new premises and extensions to existing premises. Governments play an active part in capital formation since they may cause 'forced' savings to be made through the imposition of taxes and they may also be able to attract voluntary savings by offering various government stocks on the investment market. They may also contribute to the process of capital allocation by investing directly in government-owned industrial or service plants and by allocating various subsidies to different sectors and industries. Most of their investment, however, is in the form of social and overhead capital projects, such as the building, maintenance and expansion of schools, roads, energy resources, social welfare services, and technological research centres; all of which are designed to increase the 'intangible' assets of society. The generation of these intangible assets is not the exclusive prerogative of governments since there are a number of ways in which individuals, firms and industries contribute to the process: money spent by a father on the education of his sons and daughters or money spent on industrial training schemes and research projects by a firm, represent additional forms of indirect investment in social capital that, through time, is destined to become an intangible asset not only of the individual or firm but also of society as a whole.

Capital formation is thus effected in various ways by different individuals and groups and for a variety of different reasons.

6.1 SAVINGS

The ability to generate savings in any system is related to the operation of a number of endogenous processes of income determination and the generation of a series of exogenous stimuli to those processes.

6.1.1 *Income*

Income is the fundamental determinant of both consumption and saving: as disposable income rises, consumption and savings levels both increase, though at rates that vary according to the prevailing level of income. At very low levels of disposable income, consumption may, of necessity, exceed income: families, firms or groups may somehow be spending more than they actually receive in direct income, and consequently they have to borrow (or beg) in order to make ends meet (for example, point A on figure 6.1). In such circumstances, all additional income that may accrue is spent on additional consumption items and on the repayment of any debts that may have been incurred, so that for a time the MARGINAL PROPENSITY TO CONSUME increases as income levels rise ('marginal' is the economists' technical word for 'extra'). This condition may in fact continue for some considerable time since once the 'bread-line' income situation has been passed, rising standards of living mean that further increases in disposable income are spent not only

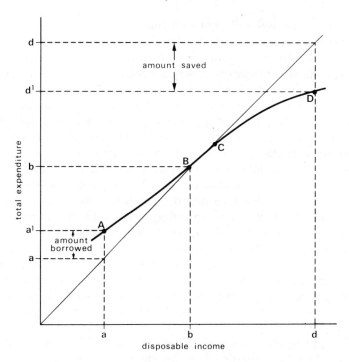

Figure 6.1 CONSUMPTION, SAVINGS AND INCOME.

on more of the basic necessities of life but to a greater extent on 'luxury' items of living. The social habit of 'keeping up with the Joneses' begins to develop as incomes increase, and is itself increasingly propagated through the 'demonstration effect' (Nurske, 1953), whereby the benefits of certain living standards are visibly demonstrated to all sections of society. Consumption expenditure and income are thus equated over a range of low to medium levels of income (from point B to C in figure 6.1). However, beyond a certain critical 'break-even' or 'take-off' point, consumption expenditure (whether on necessities or on luxuries) may not use up all available income, and saving becomes an increasing possibility. In fact, each successive increase in income beyond that point leads to a proportionately larger increase in the amount of that additional income which is saved, or in economic terms, the MARGINAL PROPENSITY TO SAVE increases (points to the right of C in figure 6.1).

Since the level of saving is primarily related to income level in this way, it follows that savings can only be increased by raising income levels. The income level of individuals, firms or groups of any kind is itself largely determined by the level of production or output that each of these groups can maintain. Output and productivity, however, can only be increased through the substitution of capital for land or labour or through increasing the scale of operation which is also dependent on the availability of capital (chapters 3, 4 and 5). Consequently, the only way in which income can be effectively raised is through increasing the level of investment that is itself dependent on the level of saving: if incomes are low, savings are low and so incomes remain low; whereas, if incomes are high, savings are high and so incomes can be increased. Equally, if savings are at a low level, incomes are also at a low level and so savings remain low and, conversely, if savings are high, incomes are correspondingly high and so savings can be increased. The mechanisms of savings and income generation thus create two different growth systems: the one is a low-level equilibrium system possessing a degree of homeostatic stability in which the endogenous (internally generated) processes are effectively income- and savings-depressing in nature, whereas the other is a high-level system characterised by endogenous processes that are income- and savings-inducing in nature.

Although the low-level equilibrium system may possess a degree of homeostatic stability, it is clear that it is not a *perfectly* stable-state system since it is manifestly possible for individuals, firms and economies to pass from that low-level state to the other high-level states in which growth can proceed. The economic development of most nations attests to this possibility, though it is difficult to assess the crucial 'break-even' or 'take-off' point at which the transition is made. On the empirical evidence of the experience of several economies it would appear that the proportion of income that must be saved (and then invested) in order to move the system from the low-level semi-stable state of equilibrium is in the order of 10–15 per cent (Lewis, 1955; Mountjoy, 1963; Rostow, 1963). It can also be theoretically demonstrated that this represents the critical minimum level of savings that must be achieved if a sufficiently regular and significant increase in income is to be secured and a resulting sufficient proportion of income is to be saved in order to continue the process of income growth. The proportion of income that has to be invested in order to secure continued growth in this way is dependent on the extent to which savings are wisely invested and also upon the overall rate of population growth that characterises the economic system. In a low-level economy of the semi-stable kind, the 'productivity of capital', as measured by the SAVINGS–INCOME RATIO, is in the region of 4 : 1 (that is, in order to secure a final income of, say, £1m, it is necessary to save and invest £4m). (In systems that have been able to make the transition from the low-level equilibrium state, the ratio is typically rather lower than this—often, in the order of 2½ : 1 or 3 : 1— because as development proceeds, and educational and technological levels increase, wiser and generally more productive investments are possible [section 6.2].) Population growth rates are often fairly high in low-level systems—rates of 1.5 per cent per annum being about the norm (section 4.1.1). If, then, the capital to income ratio in a given system were 4 : 1 and the population growth rate were 1.5 per cent per annum, it would be necessary to invest 6 per cent of the total income of the system in order merely to maintain present standards throughout. (This is because if 6 per cent of the national income were invested, and if the rate at which that investment was converted back to income were 4 : 1, there would be a growth rate of 1.5 per cent in national income: the population increase of 1.5 per cent would, however, completely absorb the additional income so produced.) In order, therefore, to secure a rise in income that would be large enough to compensate for population growth, investment must increase very significantly. A 12 per cent investment rate, for instance, would lead to an income increase of 3 per cent which would imply a *net* return (after allowing for population growth absorption) of 1.5 per cent; by the same token, an investment of 15 per cent would lead to a total *net* return of 2.25 per cent increase in *per capita* income. Either of these rates would be sufficient to guarantee future investment at the same level. Both theoretically and empirically, therefore, it can be seen that the transition from a low-level system characterised by low income and high population to a system in which a regular and significant rise in *per capita* income is sustained, requires that the proportion of national product invested increases from somewhere in the vicinity of 6 per cent to a level of between 10 and 15 per cent.

6.1.2 *Stimuli*

Since the endogenous mechanisms of the low-level system tend to be income- and savings-decreasing, it is clear that the stimulus necessary for this change in savings level to occur must be exogenously generated (that is, from outside the system), at least initially. Furthermore, the size of the stimulus must obviously be considerable. This does not mean to say that the stimulus must necessarily be of a 'once and for all' nature: the repeated and successive application of numerous limited stimuli will have a cumulative effect and the critical 'take-off' level will eventually be reached, albeit over a longer period of time (Liebenstein, 1957).

The main stimulus must inevitably be one that changes the population's attitudes towards both existing conditions of life and income levels. The *idea* that saving can ultimately lead to greater overall prosperity must be diffused and then accepted by all sections of the community. The process of the diffusion of ideas is dependent on the existence of good communications: the greater the contact-frequencies of the receivers or adopters and the greater their educational background, the more quickly are ideas diffused, and, significantly in the present context, the more likely are they to be accepted (sections 2.1.1 and 6.2.2). This represents the first obstacle to the process of stimulating the system, since contact-frequencies, educational levels and communication channels are all poorly developed in low-level systems. Under such circumstances, the stimulus can, at best, be only gradually effective as it may take some considerable time for the idea that saving is essential to reach either the whole community or a sufficiently large number of people. Furthermore, although the idea may eventually be diffused to all sections of the community wherever they may be living, the proportion of the population that might actually be able to put the idea into effect is likely to be very small, not simply because of their generally poor level of education but mainly because of their low level of income. With income levels very near to the 'bread line', the possibilities of persuading individuals to save are limited.

Yet even in desperately poor economies there are *some* individuals in the community, such as landlords or merchants, who are relatively wealthy and, potentially at least, there are some expenditures that might be diverted into savings by nearly all the members of the community. The 'leaders' of most underdeveloped communities usually live at a level that is comfortably above the basic subsistence level and often they spend quite lavishly on such items as entertaining, employing large numbers of servants or building mansions, palaces and monuments of one kind or another. The ordinary working members of the community, usually peasants, are also able to save enough to spend on various social events, customs and ceremonials such as funerals, weddings or initiation ceremonies; it was estimated, for example, that some 7 per cent of rural expenditure in India went on such purposes in the days before development began (Wolf and Sufrin, 1955). If the idea could be engendered that these kinds of expenditure could be more gainfully employed (in savings and investment), a start might be made towards stimulating the economy in an upward direction of growth.

Some form of 'cultural revolution' appears to be necessary if attitudes are to be changed and savings increased. But the acquisition of new attitudes is of little use unless it is accompanied by a marked increase in the availability of savings institutions within the community's territory. In order to get individuals and small enterprises to save, there must be available as wide a range of savings institutions as possible at a local level, since experience shows that the amount of saving achieved in any system is correlated with the availability of post offices, friendly societies, building societies, stocks and shares brokers, credit institutions, banks and all other forms of financial institutions. The geographical diffusion of these institutions to a very localised level is consequently an essential part of the drive to increase the endogenous supply of savings.

The existence of adequate financial institutions is, however, of no consequence unless the community becomes confident that its savings are safe when entrusted to them. The early history of banking in most of the developed countries of the world reveals that the development of confidence may be a difficult and lengthy process in which many 'failures' may be inevitable. For some largely unknown reason—but presumably one based on false information received from the spread of rumours—confidence in any institution can be shattered overnight and a 'run' on the institution may ensue. Once this happens, the institution may have to suspend payments in order to protect its reserves, but the effect of such a move is likely to aggravate the situation and to further the lack of confidence that gave rise to the situation. In such circumstances, the institution may well 'fail' unless confidence can be restored very rapidly. Nor does this possibility apply merely to the early development of financial institutions, though it is then that it is most likely to occur; not infrequently in modern, highly developed economic systems, confidence is upset from time to time, thereby causing runs not only on the funds of individual firms but even, in extreme cases, on the currency of the nation.

Many economists believe that the savings that might be generated in this way would not be sufficient in themselves and that they must be supplemented by 'the emergence of a new class in society which is more thrifty than all the other classes . . . and whose share of the national income increases relatively to that of all others' (Lewis, 1955, p. 226). This class may comprise either individuals and groups whose primary motivation is to make profit, or governments whose primary motivation is also to eke out a percentage 'profit' from the national product through taxation and various other means (sections 9.4.1 and 9.4.2). The process of 'converting' individuals or groups in a low-level economic system to 'profit makers' would be lengthy, difficult, or limited if not actually impossible for the reasons already discussed. Initially, therefore, this new class may have to be imported into the economy, and as such represents an exogenously generated stimulus which, in the experience of most developed countries of the world, has been in the form of foreign traders and foreign investors (African Asians, for example). The 'demonstration effect', by which the individuals in the community begin to recognise and feel the benefits that this small group can bestow both on themselves and on the community, eventually leads to a quickening of the pulse of saving, since incomes begin to increase. The government of the nation or state may then find that it too can play a role in the 'demonstration effect' by introducing various savings and investment policies that may be either financially based or more indirectly based on the re-organisation of social and employment structure (sections 9.4.3 and 9.4.4).

Nevertheless, it is doubtful whether changes in attitude and the emergence of the profit-making entrepreneur are together sufficiently potent to create the required 10–15 per cent rate of capital formation, since both are based on raising endogenous savings that are in short supply. In such a situation, the only remaining practical alternative is to search for external help in the form of loans or gifts from other, presumably developed, economic systems abroad. The principle underlying this possibility is that if a sufficient amount of capital either in the form of real money or in the form of tangible or intangible assets can be borrowed for a period long enough to get the economy out of the low-level equilibrium trap, the amount borrowed can be repaid from the savings and additional income that will result from the capital so invested. Certainly, this mechanism has been widely employed by many countries in their attempts at economic development (table 6.1). Britain, for example, relied on Dutch capital in the seventeenth and eighteenth centuries and, in turn, having made the transition from underdeveloped to a developing state, financed (and continues to finance) the development of many territories in the world. Foreign capital may thus constitute a further exogenous stimulus to economic development at a national level, and by the same token regional development and the development of individual firms and industries may be stimulated by, if they are not actually dependent upon, borrowed investment from external sources such as national governments (as in the case of regional development) or other firms and industries. Borrowing is a widespread practice that is essential not just for the initial stimulus that it may provide but also for the maintenance of the critical minimum investment level, so that 'cases where virtually all capital has been obtained from internal sources are not typical' (Mountjoy, 1963, p. 97). There are, nevertheless, many problems involved in trying to persuade investors to allocate their funds to projects in systems that appear to be in the low-level equilibrium trap (section 6.2.1), and there are equally many government controls on the international mobility of capital (section 9.4.4).

6.2 INVESTMENT

Industrial and private entrepreneurs are normally concerned to maximise profits or to minimise costs and risks, both of which relate to their perception of the alternatives that are considered to offer the most 'satisfactory' outcome in any situation (section 2.3.4). With such principles guiding their basic motivation, private investors are most likely to allocate their capital to such projects as appear to offer safe forms of investment or to those in which the return is thought to be sizeable. The result is that private capital tends to flow to already developed sectors or regions of the economy and consequently it may be argued that 'the market mechanism fails to achieve an optimal allocation of resources' (Richardson, 1969, p. 191). For this reason, the government of any nation or region may find it necessary to allocate *its*

TABLE 6.1 NET FLOW OF FINANCIAL RESOURCES TO DEVELOPING COUNTRIES, 1960, 1965 and 1970 (in million US $). Source: O.E.C.D. (1971): A–types of capital flow; B–1970 contributions of member countries of DAC (the Development Assistance Committee of the Organisation for Economic Co-operation and Development [OECD]).

A *Capital Flows by Type*

	1960	1965	1970
I *Official Development Assistance*	4665	5916	6808
1 Bilateral grants	3692	3714	3398
2 Bilateral loans	439	1854	2386
3 Contributions to multilateral institutions	534	348	1124
II *Other official flows*	300	283	1159
III *Private flows*	3150	4121	7575
1 Direct investment	1767	2468	3408
2 Bilateral portfolio	633	655	809
3 Multilateral portfolio	204	247	343
4 Export credits	546	751	2174
Total net flow	8115	10 320	15 542

B *Capital Flows by Country of Origin* (1970)

Country	Value of flows ($m)	Country	Value of flows ($m)
Australia	384.9	Japan	1823.9
Austria	96.1	Netherlands	456.6
Belgium	308.5	Norway	66.7
Canada	626.4	Portugal	64.7
Denmark	97.2	Sweden	229.3
France	1808.3	Switzerland	137.4
Germany	1487.0	UK	1258.7
Italy	724.9	USA	5971.0

available capital in such a way as to counteract the normal mechanisms of the investment market. Perhaps the most significant aspect of investment is that it is primarily responsible for technological innovation in the system. Technological innovation is a crucial correlative of investment and, indeed, together innovation and capital constitute the fundamental elements that many observers consider to be the *primum mobile* of the system (Dewhurst, 1955; Kindleberger, 1965; Mears and Pepelasis, 1961).

6.2.1 *The market mechanism*

The extent to which savings may be invested in any economic system or in any sector or location within that system is primarily related to the confidence of investors and their attitudes to risk. All investments involve a degree of risk, for there can never be any *certainty* that a project in which an investment is placed will be entirely successful or, indeed, that the return on the investment will be of any given magnitude.

Individuals react very differently from each other in risky situations, but there can be be little doubt that the size of potential gain is significant in determining investors' attitudes: provided that the pay-off is high enough, even a moderately cautious person may occasionally take a reckless gamble (section 2.2.3). The more an investment appears to offer the prospect of a sizeable and short-term reward, the more likely it is that investors will be attracted to it. For this reason, investment funds tend to flow more readily to projects or areas in which the expected productivity of capital is highest: thus, funds are more easily attracted to developed economies and regions in which the capital to output ratio is low than to low-level semi-stable systems in which the ratio is high. Similarly, it is because investment in social and overhead capital schemes generally produces only moderate financial returns over very long periods of time, that relatively few funds are attracted to this kind of project from the private sector of the investment market. Indeed, it is precisely in this sphere that the market mechanism appears to fail almost completely: with some notable exceptions, the provision of adequate housing, education, medical and welfare services, sanitation, street lighting and *long term* projects designed to create an improved resource base, or merely a pleasant environment in which to live, is rarely sponsored by capital allocated from the private sector. Yet it is upon the development of these kinds of services that the long-term prosperity and living standards of any system are ultimately dependent.

In the absence of high or quick returns, most investors are likely to be attracted to *safe* investments. Indeed, in some cases this kind of investment may be preferred to all others because, like investment in building societies, there is little risk and the funds are 'as safe as houses'! Such conditions are most likely to apply to projects similar to ones that have already succeeded elsewhere, and to those sponsored either by well-established individual firms and industries with good records of growth and share dividends, or by regions and countries with similarly good records. It is very often the case that investment in underdeveloped regions and in new industrial ventures is regarded as particularly unsafe: far from creating favourable conditions for investors, many developing regions, nations or firms may, perhaps unwittingly, create so many clauses and conditions of investment that they actually tend to frighten investors away. The ever-present threat of state nationalisation of tangible assets developed by outside interests is often claimed to be one of the factors tending to deter direct foreign investment in the developing nations of the world.

The extent and the reliability of the information relating to the alternative opportunities available, may be a further factor affecting the operation of the investment market. Other things being equal, the more that is known about any project, and the more reliable the information is thought to be, the greater the confidence of the investor is likely to be. Most investors' contacts tend to be within existing industrial conurbations or at least within a developed region, so that there is a greater likelihood of their receiving relevant information about investment opportunities in their own already developed environment than about those that may exist elsewhere. (A distance–decay function characterises investors' information fields in just the same way that it characterises any other information field [section 2.1.1].) Furthermore, information relating to investment possibilities is usually much more readily available for existing firms, industries and regions than it is for others.

All things considered, therefore, investors are most likely to be attracted either to 'safe' propositions—ones in which the return is thought to be sizeable—or to those in which they can place the greatest confidence consequent upon the reliability and extent of their sources of information. The net result is that investment funds tend to be channelled towards existing sources of income or capital and away from newly-created sources or ones that are stagnating. The biblical suggestion that 'to every one who has will more be given, but from him who has not, even what he has will be taken away' (Matthew 25: 29), seems to describe the mechanism succinctly! The process of cumulative causation (section 3.2.2) is thus further intensified as capital flows continue to be attracted towards expanding firms and regions at the expense of those firms and regions that may either be caught in the low-level equilibrium trap or are merely beginning to decline economically. Certainly, there is ample evidence to suggest that there is a widespread pattern throughout the world of 'a net capital transfer from lagging to growing regions' (Keeble, 1967, p. 262; figure 6.2), the effect of which may be to cause widespread and increasing regional disparities in income and prosperity. In such circumstances, further social, human and economic problems arise. Seeing the prospect of greater wealth in some developing region, people from surrounding areas are likely to leave their original homes and move in, permanently, to the expanding town or city at the centre of the development, thereby further polarising development into a few localities and causing negative backwash effects that bring economic stagnation or decline to the surrounding areas.

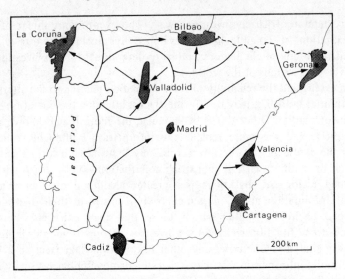

Figure 6.2 CAPITAL FLOWS IN SPAIN capital flows to the agglomerating areas of Madrid, La Coruna, Bilbao, Valladolid, Gerona, Valencia, Cartagena and Cadiz. Source: Lasuen (1962).

In addition to creating regional disparities of growth and income, the normal mechanisms of the investment market also tend to create disparities in growth rates between different sectors of the economy, so that the total structure of the national economy may become unbalanced. Too great an investment in certain industrial sectors at the expense of development in agriculture (for example), may lead to an eventual slowing down of the growth rate of the originally highly capitalised section. This is simply because as industry advances, demands for food are increased due to the increased wealth of industrial workers, and if those demands cannot be met because the agricultural sector has not been sufficiently expanded, the whole economic system begins to collapse. This, at least, has been the experience of several underdeveloped countries in their attempts to promote economic growth through industrialisation.

Because funds are generally attracted to short-term projects the normal market mechanism does not guarantee either sustained or adequate growth in the long term. Rather, overall economic growth rates tend to be very variable simply because investment levels are determined by short-term changes in business prosperity or confidence.

In short, the mechanisms of the investment market may not lead to an optimum allocation or distribution of resources, income or prosperity either between different sectors or between different locations in the system: nor can they guarantee stability of economic growth.

Capital

For these reasons, the basic objective of public investment must be to supplement these mechanisms by adjusting the allocation of investment in different sectors of the economy, redistributing income and wealth between sectors and geographical locations and by stabilising economic activity generally. Patently, additional measures such as taxation and legislation may be necessary, but it is through public invesment in the form of subsidies and grants towards the maintenance or creation of plant, equipment and services that progress in each of these three spheres may be achieved (section 9.4).

6.2.2 *Innovation*

Technological innovation is one of the main end products of investment, and it is through its application to different elements of the production process that change is induced in the system. A change in any one element leads to further changes in other elements since, within any system, all the elements are interrelated. This effect can be demonstrated theoretically by means of a four quadrant graph that shows the relationship between land use, output, labour productivity and labour inputs at various levels of technology (figure 6.3A). All possible alternative crop combinations (between two commodities) are described in the upper right quadrant, the necessary labour inputs associated with each combination are shown by the line in the lower right quadrant, while labour productivity and the resulting total output are shown respectively in the lower left and upper left quadrants. Any increase in labour productivity consequent upon the adoption of some technological innovation

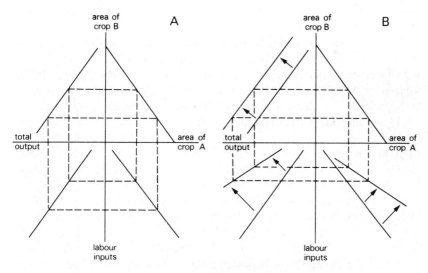

Figure 6.3 PRODUCTIVITY CHANGES Source: Inukai (1971).

has the effect of shifting the input–output curve upwards, and as this happens shifts also take place in the curves relating output to land use (upper left quadrant) and labour input to land use (lower right quadrant). A new land–labour–output system emerges (figure 6.3B) in which both land and labour productivities are higher than they were in the system typified by a lower technology (Inukai, 1971). A sizeable difference exists, however, between that which may be technologically *possible* in any economic system and the level of technology *adopted* either by the main firms or by the economy as a whole. As a general rule, it would appear that certain firms become innovational leaders and adopt new techniques well before the majority of other, usually smaller, firms in the same industry, but even the leaders operate at a somewhat lower technological level than that which may be theoretically feasible (Brozen, 1950) because it is not always possible to adopt the very latest techniques as soon as they become available.

The pattern of adoption proceeds in a manner that is similar to that of the diffusion of information (section 2.1.1): initially the number of adopters is rather limited, but through time the pace of acceptance quickens and follows the general form of a logistic S-shaped curve. Such a pattern of adoption has been shown to occur in several activities (Mansfield, 1961), though the actual rate of adoption varies considerably from activity to activity. Whereas it took about 15 years for 50 per cent of all pig-iron producers in the USA to adopt the coke oven, continuous mining machines had been adopted by 50 per cent of all coal producers within 3 years of their first introduction into the industry. Similarly, 50 per cent of all British farmers had adopted the technique of artificial insemination in their dairy herds within about 8 years of its being first introduced in 1944, but it took 18 years to reach the same level of adoption of combine harvesters which were first introduced in 1928 (Jones, 1962). Such patterns characterise the adoption of new techniques in a wide range of social, cultural and commercial activities (Hägerstrand, 1952; Kniffen, 1951A/B and 1965; Zelinsky, 1967; figure 6.4). The 'primary', 'diffusing' and 'condensing' stages of the spatial diffusion process (section 2.1.1) are all evident. Hägerstrand (1952) has shown, for instance, how the level of adoption of cars in Scania between 1918 and 1930 varied spatially from the innovational centre of Malmö. The 'primary' stage occurred between 1918 and 1920 when the greatest adoption took place near Malmö, and then from 1922 to 1926 the pattern corresponded to the 'diffusing' stage with the greatest increase in adoption occurring at some distance away from Malmö. The 'condensing' stage occurred, finally, after 1928. Because of such time lags, the rate at which innovations are adopted by all people living in each location varies with distance from the innovational centre, and correspondingly therefore, the S-shaped adoption curves of each locality take a slightly different form according to whether the locality is progressive (and an

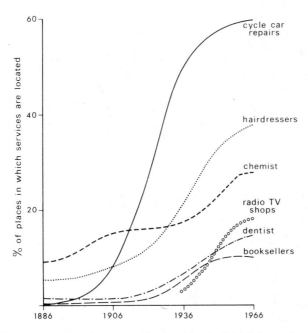

Figure 6.4 DIFFUSION OF SERVICES Nord and Pas de Calais Départements, France.

early adopter) or backward (and a late adopter; see figure 6.5). In terms of the adoption of cars in Scania, the Malmö region was the most progressive, whereas northeast Scania (further away from the innovation centre) was the least progressive. The time lag between the most progressive and most backward regions appears to be greatest during the 'diffusing' stage of adoption (between the 50 and 60 per cent adoption level): a feature that Jones (1962) has also revealed in his study of the diffusion of T.T. milk production in England and Wales. Interregional time lags such as these are due to a number of characteristics related to the innovation itself, the industry within which it is introduced and to the channels of communication through which information about the innovation is diffused.

The pace at which information relating to a particular innovation is diffused is related to the existence of appropriate media of communication by which all potential adopters may be informed of its characteristics and availability. The contact-frequencies of the individual decision maker are thus fundamental in determining whether and when he receives the necessary information (section 2.1.1). Such information may be either public or private and transmitted to the potential adopters through the mass media or by person-to-person contact. In general it seems that as far as innovation is concerned, the leaders gain their information from scientific institutions and

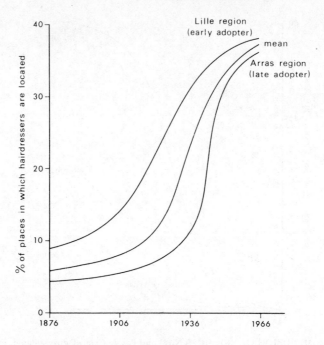

Figure 6.5 EARLY AND LATE ADOPTERS diffusion of hairdressing salons 1876–1966. Nord and Pas de Calais Départements, France.

their publications, their own research establishments or from various educational courses (including 'refresher' courses), whereas the majority of the potential secondary adopters are dependent on person-to-person contact or the mass media, both directly through specialist radio and television programmes, and indirectly through more 'popular' programmes such as 'The Archers' or 'Tomorrow's World' that often contain a great deal of information (not to mention indoctrination) about modern technology. In other words, relevant information 'trickles down, usually in an increasingly simplified and more relevant form, from the original sources to the laggards' (Jones, 1962, p. 395).

To receive information is, however, a vastly different proposition than it is to act on it. Decision makers, irrespective of any technical or other considerations, vary considerably in their ability to adopt. Indeed, it would appear that the probability of adoption is dependent on the industry in which the innovation is introduced, for decision makers in some industries seem to be more able or more prepared than others to undertake innovation.

Within a given industry the probability of adoption is likely to be highest in those firms that are experiencing 'boom' conditions and that are consequently expanding at a relatively rapid rate (Mansfield, 1961). Similarly, firms that are hoping to increase their share of the market will also try to adopt the latest available techniques so that they are productively more efficient than their competitors. But it is usually large scale firms that are most able to take advantage of new techniques and it is precisely these firms that are able to afford extensive research departments of their own to develop better and better production technology. In agriculture, too, the speed and degree of adoption are positively related to farm size and income, which, in turn, are positively correlated with high educational attainment and socio-economic background of the farm owner. A highly significant relationship between the 'the level of adoption of a number of innovations and farmers' total gross income, socio-economic status, the use of information and advisory services, farming type and farm acreage' was, for example, established in a study of farms in mid-Cardiganshire (Jones, 1962, p. 398).

Decision-makers in large establishments thus appear to be faster in adopting new techniques than those in small firms. Since advantage and scale are cumulative phenomena (see chapter 3), innovation is also cumulative, and innovational leaders are increasingly to be found in large scale firms. This is not to deny the possibility of independent small-scale operatives discovering new techniques, but it is clear that their possibilities are proportionately reduced as amalgamation proceeds and produces larger and larger operating units. Scale and innovation are mutually interrelated phenomena.

The technical and economic characteristics of innovations have a considerable bearing on their potential adoption by users and it is precisely because different decision makers regard different innovations as either technically too advanced or economically unjustifiable that there is such a considerable difference between that which is theoretically feasible technologically and that which is generally adopted.

The technical attributes of innovations involve considerations of their complexity, divisibility, compatibility and conspicuousness. The more *complex* an innovation, the fewer is the number of its adopters since complexity necessitates greater skill or ability either in comprehension of the technique or in its use and maintenance. Similarly, the extent to which an innovational practice is *divisible* may affect its rate of adoption. If a technique can be given a small-scale try-out, or if the innovation is a small one, its chances of adoption are greater than if it is indivisible. For instance, if a large piece of equipment has to be introduced totally and irrevocably without a trial period, any potential adopter would tend to be rather more wary of it and would probably seek advice from other people who had already installed it. Such searching around for information would, of course, delay still further the adoption of the machine. In many ways it would seem obvious that large scale firms might possess some advantage over smaller firms in this respect, since they are better able to stand any failure which may be incurred. Yet it can equally be argued that to the large firm risks are even greater than to the small firm, and that they therefore possess no apparent advantage. The *compatibility* of an innovation to existing methods is also related to the problems of complexity and divisibility, since if the technique were compatible with (or similar to) existing methods, it could be more readily assimilated by potential users (Brandner and Straus, 1959). Finally, *conspicuousness* may sometimes be an important factor in speeding adoption, especially if the innovation carries a certain prestige value with it. The rapid diffusion in the consumer market of colour television sets, dishwashers and deep-freezers may well be due, in part at least, to the fact that they are effectively social status symbols.

The *economic* considerations of an innovation that have to be taken into account include its characteristics of cost and effectiveness, which in turn have to be viewed in terms of the cost and effectiveness of present techniques. The more profitable an innovation is likely to be in terms of the increased production or profit it might create, the greater is its likelihood of adoption. Once again, however, if the innovation represents a sizeable investment and requires a large capital outlay, its probability of adoption is decreased. Furthermore, it is important to recognise that no matter how profitable an investment a particular innovation might appear at any given time, its adoption will hardly be economically justified so long as *existing* plant or machinery are relatively new or yet have a considerable life expectancy. Mansfield (1961) has suggested that in the USA, the relationship between profitability (P), size of investment (S) and the rate of diffusion (B), takes the general form

$$B = E + 0.53P - 0.033S$$

where E is a coefficient that varies from industry to industry.

It appears, therefore, that economic considerations of profitability and the required scale of investment help to determine the rate at which innovations are diffused both organisationally and spatially.

FURTHER READING

Chappell and Webber (1970); Cole (1960 and 1962); Dinsdale (1965); Gwilliam (1970); Lewis (1966); McBride (1970); Manners (1964); Myint (1964); Schultze (1964); Weinberg, M. A. (1971).

7
Transfer

In a landscape system where locations are geographically separated by distance, the transfer of goods, materials and individuals is an inevitable necessity of spatial organisation, since both producers and consumers are located at different points in geographical space. Considerations of transfer consequently constitute a further element that has to be considered by the entrepreneur in the locational decision-making process.

The assembly and distribution of raw materials and products involves a cost that results both from the 'direct' charges that are levied by transport carriers, and from a series of more 'indirect' charges that arise from the variations in travel time, safety, reliability, accessibility and flexibility that characterise the different transport media. The relative importance of such costs, like all others, varies from product to product, from firm to firm, from industry to industry and from individual to individual depending on the total cost structure involved; yet it appears that 'compared to other costs, the transportation cost of most products is a relatively small part of the total' (Norton, 1963, p. 149; table 7.1). It is similarly true that transfer costs have been gradually reduced relative to other costs over the last hundred years or so as a result of innovations in transport technology. Yet despite these two characteristics, transfer costs remain locationally significant for all forms of economic and human activity, mainly because transfer over great distances normally incurs higher charges than transfer over short distances. Transfer charges thus exert a 'frictional' or constraining influence on movement—a characteristic further accentuated by the entrepreneur's need to 'minimise' transfer charges in just the same way that he attempts to minimise all other costs involved in the production process.

7.1 Direct Costs

The charges made by carriers for the transport of any commodity represent to the entrepreneur the direct costs of transfer and tend to reflect the

TABLE 7.1 TRANSPORT COSTS: production industries, Britain 1958–68. Sources: Census of Production 1958, 1963 and 1968 (HMSO).

Industry	Cost as % of Net Output		
	1958	1963	1968
Mining and quarrying	26.04	26.15	27.14
Food, drink and tobacco	13.62	13.94	12.86
Coal and petroleum products	11.74	17.23	13.14
Chemicals and allied industries	6.98	7.03	7.21
Metal manufacture	6.83	6.64	6.77
Mechanical, instrument and electrical engineering	2.65	2.68	2.73
Shipbuilding and marine engineering	1.30	1.22	2.14
Vehicles	2.16	2.16	2.34
Textiles	3.28	3.34	3.56
Clothing and footware	2.40	2.42	2.46
Paper, printing and publishing	5.68	5.73	6.21
Construction	6.65	6.67	6.71
Gas, electricity, water	2.81	2.74	2.01

carriers' costs of production and conditions of demand. Most carriers adopt the principle of 'charging what the traffic will bear', by which the scale of charges is related to the physical characteristics of different commodities and to the distances over which they are to be transferred (Locklin, 1960; Norton, 1963; Taafe and Gauthier, 1973).

7.1.1 Distance

The carrier is faced with two different, though mutually interrelated, sets of costs in providing transport over any given distance. Before any journey can be made, tracks (roads, railways and canals), rolling stock (lorries, engines, wagons, carriages) and terminal equipment, including clerical and administrative staffs, have all to be provided. Such overheads constitute a series of FIXED COSTS that have to be met from the charges levied to the users of the services provided. The more users there are, the less, *proportionately,* is the share of fixed costs that any one user must bear, and consequently the fixed cost element of the transfer rate is reduced up to a point (figure 7.1A). This possibility presents the carrier with a dilemma: attractive (low) transfer rates can be generated only by increasing the number of users, yet this may only be possible by offering attractively low transfer rates. While it may seem feasible in theory to offer low rates in the hope of attracting sufficient custom to cover its proportional share of the fixed costs, there is no guarantee that this will be the result in reality. Carriers for whom fixed costs form a relatively

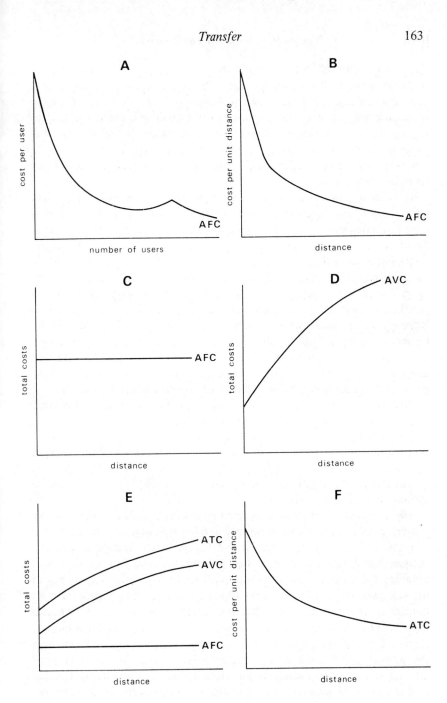

Figure 7.1 CARRIERS' COSTS *Key* AFC, Average Fixed Costs; AVC, Average Variable Costs; ATC, Average Total Costs.

large proportion of total operational costs (railways, for example) are particularly susceptible to this problem. Offers of cheap day excursions or of generally low rates on unremunerative lines may, to the layman, seem to be the logical way of attracting custom, but in fact there is ample evidence to suggest that in many ways transport demand is essentially inelastic and that lower prices do not necessarily create increased demand (section 8.1.1). In this way, fixed costs fundamentally affect the commercial viability of any transport undertaking.

It is also important to recognise that fixed costs have to be met irrespective of the distances over which different users may wish to transport their commodities. The cost per kilometre does, of course, decrease because the fixed cost is distributed over a greater number of distance units (figure 7.1B), but as far as the carrier is concerned the *total* fixed costs are essentially the same no matter what the distance involved. For example, the fixed costs of provision and maintenance of the railway line between Thetford and Swaffam in East Anglia amounted in 1962 to £20 100 per annum (British Railways Board, 1963); since the line had to be maintained over the full 37-kilometre length in order to allow transfer between Thetford and Swaffam, the same fixed costs were bound to apply to any journey between intermediate stations only. Graphically, therefore, the relationship between fixed costs and distance is represented by a horizontal line (figure 7.1C).

The VARIABLE COSTS of transport (sometimes known as the 'line-haul' costs) are those that are directly incurred by the journeys made, and as such they include the costs of fuel, oil and any other materials used *en route*. Since consumption of such materials tends to be heaviest on short distance journeys, variable costs per unit distance do not increase in direct *pro rata* proportion to the length of journey involved though they do increase in absolute terms (figure 7.1D).

Together, the characteristics of fixed and variable costs are such that the carriers' *total* costs, while increasing *absolutely* with distance of transfer, tend to increase at a relatively declining rate (figure 7.1E). Translated into user charges, this situation gives rise to high unit transfer rates for hauls over short distances and progressively lower unit rates for long-haul transfers by any of the transport media (figure 7.1F). Tapering rates are thus typical of most transfer charges and directly reflect the increasing economies to be gained from long-haul transfers. Because of this, the charge for making a *through* journey between two places is normally considerably less than it would be if the journey were made in two stages, with a break in transit at some intermediate point. For instance, the single bus fare from Exeter to Tiverton (a distance of 24 kilometres) is 25p; the fare from Exeter to Silverton (an intermediate point at a distance of 12 kilometres) is 15p, and that from Silverton to Tiverton (12 kilometres) is 15p; the combined fare of this broken journey would thus be 30p—5p higher than the through rate. This practice may be modified by some transport authorities so as to allow 'fabrication in transit'. Such a concession is of particular benefit to a firm that is located at an intermediate point between its raw material supply points and its markets. The 'through rate' that would exist between the two extreme locations contacted by the firm is allowed to apply to the two separate parts of the haul (raw materials to factory, factory to market). In America, such in-transit privileges are most commonly applied to grain production (milling-in-transit), lumbering and iron and steel production; in the early part of this century it was estimated that some 300 commodities were the subject of such agreements (Locklin, 1960).

A further modification is not infrequently made to the standard distance—charge relationship by the introduction of 'blanket rates'. A uniform charge is made for transit to all locations within a given area; this may be quite small, as in France where the same passenger rail fares often apply to four or five stations within 10 kilometres of each other; or it may be considerably larger, as is the case in America where blanket rates commonly cover all locations within an 80-kilometre radius of each other and occasionally even apply to two or more states. The main advantage to the carrier is, of course, that it makes charging simpler and easier to administer, while for the user the effect may be to reduce the 'frictional effects' of distance quite considerably. The introduction of such variations to the distance—charge relationship has the effect of modifying the geographical distribution of transfer rates: patterns of equally spaced concentric isophores (lines of equal cost) that would result if charges were directly proportional to distance are far less typical of reality than patterns of apparently meandering and grouped isophores that result from the existence of tapering and blanket rates (figure 7.2; Alexander et al., 1958).

Marked differences between the alternative transfer media in terms of the composition and level of their total costs tend to establish critical distances over which each of the alternative forms appears to possess a relative cost advantage. Road transport is generally able to offer the lowest short-haul rates by virtue of its relatively low fixed costs; rail transport becomes most competitive over an intermediate range of distances, while water transport can offer the lowest rates for transfer over great distances mainly because its variable costs become relatively smaller over such distances. The relative costs of transport are constantly being modified through technological innovation, with the result that the relative advantages of different transport forms over different distances are also being constantly altered. In particular, the advantage of road transport appears to be increasing at the present time over successively longer distances, thereby causing problems of substitution and the need to subsidise certain forms of transport (section 9.5; Sharp, 1965).

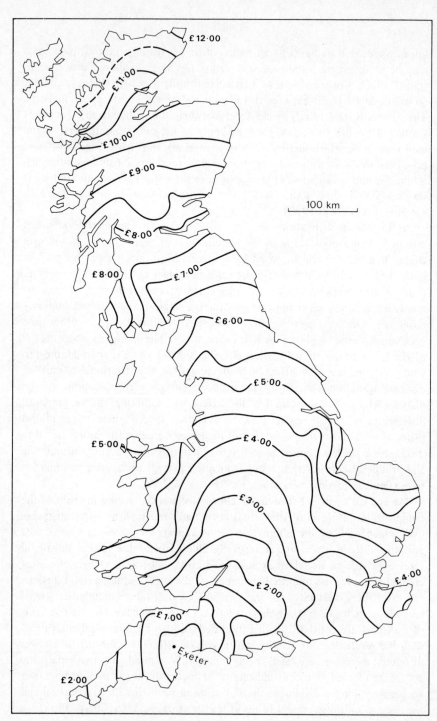

Figure 7.2 GEOGRAPHICAL STRUCTURE OF FARES isophores (lines of equal cost) of passenger rail fares from Exeter.

7.1.2 Commodity characteristics

The rate charged for the transfer of any commodity is related not only to the distance over which the transfer is to be made, but also, and in some cases more importantly, to the physical characteristics of the commodity concerned. Foremost among these characteristics are the weight, volume and value of the commodity. Generally speaking, carriers find that traffic of high value in relation to its weight can absorb higher charges more readily than bulky traffic of low value. This possibility arises because the transfer charge for goods of high value per unit weight represents a relatively smaller addition to the total cost of the delivered article than it does for goods of low value per unit weight. In other words, the elasticity of demand for the transfer service is less for goods of high value per unit weight than it is for those of low value per unit weight. In turn, this makes it possible for the carrier to make such commodities bear a higher proportion of the fixed costs of operation and ultimately to 'subsidise' the transfer of other commodities that could not bear such high charges. Railway charges in Britain were based very much on this principle, at least until 1953; the Railways Act of 1921

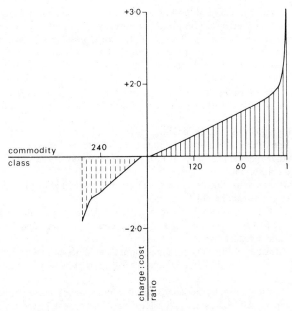

Figure 7.3 RAILWAY CHARGES ratio of charge to cost of transfer by rail of different commodities (USA). Charges for commodities 1–180 are greater than the actual cost to the carrier, whereas charges for commodities 180–260 (US standard classification) are less than the actual cost to the carrier (indicated by the dashed lines).

TABLE 7.2 GENERAL RAILWAY CLASSIFICATION: sample list of articles in each class.

Class	Commodities
1	Ashes and cinders for road repair or waste tips; chalk in bulk; coal pyrites; iron ore.
2	Basic slag; chrome iron ore; cake breeze; gannister; sand in bulk or sacks.
3	Lime in bulk; magnesite, burnt; roadmaking and road-repairing material, tarred.
4	Coal-tar pitch, in owners' tank wagons, 8 tons per truck; creosote, in owners' tank wagon.
5	Building bricks; concrete blocks and slabs; brewers' grains, wet; iron and steel scrap; manure, other than stable or farmyard; sugar beet, 6 tons per truck.
6	Ammonia sulphate, 6 tons; china clay; ferro-manganese; blooms, billets, bars and ingots, iron and steel; naphthalene in owners' tank wagons.
7	Stone, roughly wrought; cement; gypsum, packed; prepared hearthstone; kelp; boiler plates and tubes, iron and steel; molasses, 15 tons, in owners' tank wagons.
8	Armour plate; bitumen; cider apples; grain, whole-flaked or ground; pig lead, nails and spikes, iron and steel; oilcake and cattle food; potatoes and vegetables for human consumption.
9	Crude asbestos; bone ash; colliery guide rods, not to take strain; emery stone; forgings in the rough, iron or steel; Fuller's earth; plasterers' laths; marble blocks, rough.
10	Bean sticks; animal charcoal; cork, granulated or dust, press-packed, 2 tons; flower pots; iron or steel mesh for reinforced concrete; earthenware tubs.
11	Ale and porter in casks or cases; tarred felt, 2 tons; ordnance; litharge; vegetable oils in casks or iron drums, 2 tons; paper in bales or bundles; soap.
12	Copper, blooms and wrought, plate bar or strip, 5 tons; electric cable, 5 tons; glue; hay and straw, 2 tons; machines and machinery, complete or in parts, 5 tons.
13	Wooden balls for throwing at coconuts; wooden clothes line posts; colliers' baskets and boxes; cockles; cotton, raw, press-packed; window-frames, iron and steel, packed.
14	Brass bolts and nuts; electric cooking stoves, 2 tons; dried fruit; petroleum jelly in casks or iron drums; cured hams; kitchen boilers; letter boxes.
15	Laundry blue; pigments in casks or iron drums; confectionery in casks or iron drums; undressed leather; mustard, yarn.
16	Torch batteries; milk churns; china in crates; crabs; dyes and colours, not dangerous; glass sheet; paper handkerchiefs in bales.
17	Machinery belting; books in paper covers; cigarette pictures; hardware, not aluminium, nickel or pewter; raw rubber; lavatory stands and basins; tinware; complete mirrors in frames.
18	Flock bedding, packed; bicycle frames; boots and shoes in cases; broom and brush heads; wooden chairs, packed not upholstered; grass growing in boxes; clothing, not silk.
19	Alabaster; animals, live; leather goods; electric bulbs; hobby horses; boilers and water heaters, copper and brass; soldiers' busbies; mixed drapery in boxes or parcels.
20	Aeroplanes, packed; amber; animals, stuffed in cases; bagatelle tables; bismuth; cigars; silk lace; furniture, packed.
21	Gold and silver, and manufactures thereof.

established a 'General Railway Classification' of twenty-one commodity groups with the lowest categories of low value articles being charged the lowest rates over hauls of different lengths (table 7.2; Sharp, 1965, p. 46). Similarly, in the USA railroad charges vary according to the commodity transported, with some categories being effectively subsidised by others (figure 7.3).

Other criteria are also considered in establishing inter-commodity rates: if special additional facilities are required for the handling of a particular product, then the cost of these specific services will normally have to be borne exclusively by that product. Similarly, the regularity and volume of transfer may have an important bearing in establishing the user charge, since if the volume of transfer is sizeable and frequent, scale economies will allow the reduction of unit transfer costs in just the same way that transfers over long distances permit relatively lower rates to be charged. For this reason, many carriers find it desirable to charge 'exceptional rates' (that is, rates not necessarily related to value, weight or consideration of specific services) for certain commodities. The railway companies of Britain, for instance, were allowed to charge exceptional rates for any commodity, provided that the charge was not more than 40 per cent below that of its appropriate 'classification', in order mainly to attract bulk orders of regular consignments. Indeed, exceptional rates are perhaps today rather less the exception than they used to be, and charges are based to a lesser extent on the weight, volume and value of the commodity alone.

7.2 INDIRECT COSTS

In addition to the actual charges levied by the carrier, a series of further indirect costs may be incurred in the process of transfer, and these may also help to determine which method of transfer is to be preferred for particular commodities and products. In some cases, the factors of transit time, safety, reliability, accessibility and flexibility may not necessarily involve additional costs, though they may make one form of transport more *suitable* than another. The adequacy of transfer facilities in any location may be a further consideration in the locational decision-making process.

7.2.1 *Transit times*

For certain commodities, transit time may determine not merely the method of transport most suitable for their transfer, but even whether transfer is feasible at all. Highly perishable market-garden produce requires rapid transfer from producer to consumer; if it is not speedily transported, it will deteriorate so much in quality that the producer is unable to command a high market price for it. Such possible losses in revenue represent one form of indirect cost that is involved in transit considerations. It may, of course, be possible to reduce market price losses by using the form of transit offering the shortest transit time, but normally this can only be achieved by incurring greater *direct* cost outlays since the cost of transfer usually increases as transit time is reduced. Alternatively, market losses might be reduced by the provision of special forms of container designed to reduce the perishability of the goods in transit. Again, however, this involves additional expense whether in terms of better packaging or the provision of refrigerated containers, so that the net effect may be just the same as trying to decrease transit time. In deciding which form of transport to use, the producer must therefore assess whether the *marginal* cost of rapid transit is more than offset by the *marginal* market price his product may command consequent upon the reduction of its in-transit time.

This principle does not apply only to perishable commodities: goods required *urgently* as special consignments are an obvious example of non-perishable goods for which marginal cost substitution of this kind may be significant. Perhaps less immediately obvious is the fact that many passenger movements are of a similar nature: where time is limited, passengers will tend to substitute the higher transit charges of rapid means of transport for the loss of time that would otherwise have been incurred. (It may, for instance, be worthwhile paying the extra cost of travelling by air for a weekend in Paris compared with the lower fare by cross-channel ferry simply because more time can then be spent in riotous living!)

As transit times are shortened, reductions can be made in the level of stocks (inventories) that the producer has to maintain in order to meet the variations that may occur in demand for his product. Since fewer units of production are tied up 'in transit' at any given time, and since the value of stocks constitutes an element of the overhead costs of the firm, a reduction in transit time may lead to a reduction of interest charges on the firm's capital assets. In this respect, the producer must again consider whether the *marginal* cost of rapid transfer is more than offset by the *marginal* reduction of interest charges: if it is, substitution will almost certainly result.

Substitution between the various transport media is thus based on the entrepreneur's assessment of the marginal costs of expected utility whether measured in terms of market prices, time or interest charges.

7.2.2 *Safety and reliability*

The possibility of loss or damage to commodities in transit is always very real, but the extent to which this consideration is significant varies according to the type of merchandise being transported. In general, it is the perishable, fragile and high-value commodities that are most liable to loss and damage, so that the reputation both of the various transfer media and the individual carriers in this respect may well help to determine which firm and which form of transport is chosen for the carriage of such commodities.

Similarly, the reliability of different carriers in terms of their advertised or promised schedules may influence the selection of a particular carrier for a particular purpose. Perhaps the most ironic example of this problem is afforded by the alleged decision of British Railways Eastern Region to distribute their timetables for 1958 by road hauliers because delivery in time for the start of the new timetables could not be guaranteed by rail! 'Guaranteed' deliveries at a certain time or on a certain day are rare, but when they do exist they inevitably entail a higher cost than delivery by a 'non-guaranteed' method.

In both cases, safety and reliability requirements may also involve indirect expense on the part of the senders of goods. Better packaging, which may protect the merchandise from either damage or loss, is available only at additional cost, and financial restitution for damage or loss can only be obtained by taking out greater insurance cover. It is significant that from the carrier's point of view, good reputations for care and reliability often become crucial, especially in situations where direct charges are roughly similar between different firms and different transfer media (Norton, 1963, p. 140), for good reputations may obviate the additional indirect charges that poor reputations would necessitate.

7.2.3 *Accessibility, flexibility and adequacy*

The possibility of transfer from one location to another depends on the accessibility and flexibility of operation of the various transport media. Accessibility is curiously related to geographical scale in that roads are generally accessible to any location, railways and canals to rather larger areas, while airports tend to serve even larger tributary areas. This fact may make necessary the transfer of goods from one mode of transport to another, a process that tends to increase the costs of transport due to the need for intermediate rehandling of cargoes.

Similarly, the degree to which the capacity of the transport unit may be changed without causing excessive additional costs may help to make certain forms of transport competitive for irregular consignments. Thus, for example, railways may find it possible merely to add on a few more trucks to an already scheduled service and thereby be able to offer a less expensive rate for such occasional small consignments than, say, a road haulier who might have to provide a special service for those same commodities.

The adequacy of transport provision at any location is related to its accessibility and flexibility and also reflects the conditions of transit time, safety and reliability. Since 'adequacy' can only be relatively subjectively assessed, it follows that it is the decision maker's perception of transport provision that is crucial in his decision-making process. The perceived adequacy of transport constitutes one of the main locational factors for almost any economic activity (Cameron and Clark, 1966). Transport provision in southwest England, for example, accounts for the greatest source of dissatisfaction with that region from the industrialist's point of view. The largest proportion of complaints (76 per cent) are related to the road system of the region, but, probably due to the relative accessibility of roads compared with rail, there is a tendency for firms either to overcome the inadequacies of the system or simply to accept the situation as it exists. Several industrialists (11 per cent) regard the lack of air services in the region as serious, but perhaps the most crucial problem arises from the perceived inadequacy of rail links. 'Uneconomic lines in the region have already been closed. This decline in facilities has caused industrialists to seek other forms of transportation and as a result has placed greater pressure on the roads while at the same time decreasing the viability of the rail network. The facts transmit themselves to the business sector and become manifested as a feeling of insecurity' (Newby, 1971, p. 191). Indeed, here, as in other locations where transfer facilities appear to be inadequate, there is considerable difficulty in encouraging any form of industrial development, for adequate transfer facilities are a prerequisite of economic growth and regional prosperity.

7.3 FRICTION

The combined direct and indirect costs of transfer are very closely related to distance since user charges are broadly based on the distance over which goods are to be transferred. As such, transfer charges may act as a kind of 'friction' against movement since greater costs are incurred by movement over great distances than by movement over short distances. Entrepreneurs of all kinds are normally well aware of this effect and are consequently concerned to make locational and other decisions 'so as to minimise the frictional effects of distance' (Garner, 1967, p. 304).

7.3.1 *Movement minimisation*

Alfred Weber (1909) demonstrated theoretically that the most efficient location for any industrial activity would be at the point where transfer costs of all kinds were at a minimum. This point would vary according to whether the transfer costs of assembling the raw materials were greater or less than the costs of distributing the products; where distribution costs were higher than those of raw material assembly, the location would tend to be near the market ('market-orientated') and where the reverse was the case, the location would tend to be near the materials ('materials-orientated'). In other words, the position of locational equilibrium, in terms of minimisation of transfer costs, is dependent upon the relative locational 'pulls' of raw materials and products which in turn are related to their relative costs of transfer.

In order to give an approximate estimate of the relative pulls of each of the raw materials and the outputs of production, Weber devised a 'material index' for the products of each industry that was calculated by dividing the weight and transfer distance of all the material inputs by the weight and transfer distance of the product. It followed that those products for which the material index was greater than 1.0 would be 'materials-orientated' (the point of movement minimisation would be near to the materials source), and those for which the index was less than 1.0 would be 'market-orientated'. This index, however, ignores the actual cost of transfer, and Haggett (1965, p. 143) has suggested that a more appropriate *orientation index* V might be calculated by deriving 'net distance inputs' measured, as Cotterill (1950) suggested, in units of cost per weight distance (for example, \$/ton/miles or £/ton/kilometres). As with Weber's index, industries for which the value of V is greater than 1.0 are said to be materials-orientated, those for which it is less than 1.0 are said to be market-orientated.

The *exact* point of least-cost location, as far as transfer is concerned, requires a rather more detailed calculation than this. Weber himself suggested that the point at which all the relative pulls were in balance could be determined by considering the locational problem as an extension of the problem encountered in physics of finding the point of equilibrium in a parallelogram of forces. Locational 'figures' could thus be drawn to indicate the relative locations of all the material inputs and the point of consumption of the product (the market figure 7.4). The movements of materials and products that would occur within this locational figure would, in accordance with the principles already described, be in proportion to the weights of those materials and products; it follows that 'these weights represent the force with which the corners of the locational figures draw the location towards themselves, it being assumed that only weight and distance determine transportation' (Weber, 1909, p. 54; Friedrich, 1929, p. 54) and that 'the location will be near the individual corners or far from them according to the relative weight of their locational components' (Weber, 1909, p. 55; Friedrich, 1929, p. 54). The solution of these locational figures could, according to Weber, be determined in either of two ways: a mechanical solution (equivalent to what is now known as a 'hardware model'; Morgan, 1967) could be found by the use of Varignon's frame (Weber, 1909, p. 226;

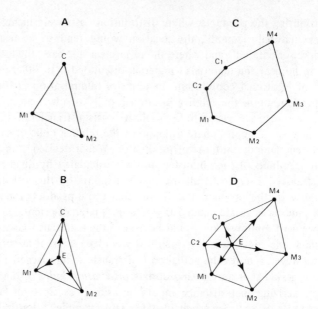

Figure 7.4 LOCATIONAL FIGURES A—locational triangle formed by the market (C) and two raw materials (M_1 and M_2); B—point of equilibrium (E) between the relative pulls of the three locations; C—locational polygon formed by two markets (C_1 and C_2) and three materials (M_1, M_2 and M_3); D—point of equilibrium (E) between the relative pulls of the five locations.

Friedrich, 1929, p. 229), or alternatively, algebra might be used to find a geometrical solution (Weber, 1905, pp. 227–40; Friedrich, 1929, pp. 229–45). Haggett (1965, p. 147) has also suggested that such 'force diagrams' can be solved graphically. An arbitrary point within the locational figure (x in figure 7.5) is selected and lines whose lengths are proportional to the 'weight' of each material or market are drawn to link the trail point with each of the material and market locations that delimit the locational figure (figure 7.5B). These lines are then joined together cumulatively to form a composite traverse which, of course, does not close (in just the same way that compass traverse surveys never initially close) (figure 7.5C). A line is then drawn to close the traverse back to the original point (x), and this line is known as the *resultant* of the traverse. The process is then repeated for a further two trial points, and the resultants of all three traverses are then plotted together on the locational figure; the point at which they intersect represents the least-cost solution of the figure (figure 7.5D). Weber argued that, under certain conditions, industrial activity might be attracted away from locations of movement-minimisation; provided that the marginal costs of substitution were favourable, agglomerative tendencies or the existence of

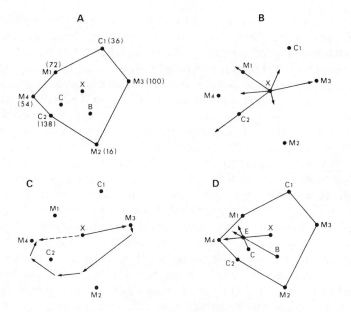

Figure 7.5 FORCE DIAGRAMS the relative 'weights' of each of the materials (M_1, M_2, M_3 and M_4) and markets (C_1 and C_2) are indicated by the numbers in brackets (based on Haggett (1965) p. 147).

a large labour supply would cause a 'deviation' from the theoretical location. In this way, if the savings in labour costs per unit output exceeded the extra transport costs, a firm would be attracted away from the 'minimisation' location to the point of labour supply.

Although the basic theory sounds attractive, it has been criticised on many grounds (Hamilton, 1967, pp. 372–4). In particular, the treatment of labour and agglomeration as 'special' inputs seems rather unrealistic, and it could be argued that had these been treated as factors of equal significance with raw materials and end products, the model would have been less 'noisy' and rather more realistic (Hoover, 1948). Nevertheless, the general theory that location is based on the minimisation of transfer costs remains essentially sound in principle.

7.3.2 *Bid rents*

Preceding and laying the foundation for Weber's theory was an analysis by von Thunen (1826) of the effects of movement minimisation on the location of agricultural land use. The basic postulate of this theory was that each plot of land would be occupied by the activity that could pay most for it, and the amount offered for each plot of land would vary according to its usefulness from the point of view of each activity—the most important determinant of this usefulness being its accessibility. On an ISOTROPIC SURFACE (section 2.1.1) the plot of land nearest to a village or farm would be the most useful for those activities that necessitated considerable transfer of either goods (for example, fertilisers) or labour from the village or farm. Farmers who wished to cultivate vegetables, for instance, could minimise their necessarily considerable movement of fertilisers and labour, by locating this form of production as near as possible to the village or farm, but in order to do this, they would have to outbid all other potential users for the use of that plot. For plots further away, involving greater movement costs, lower 'bids' would be made. In this sense, the farmer would substitute the cost of the rent of the land for the costs of transfer and, provided that the rents were lower than the transfer costs they replaced, the total costs of production would be reduced. The ability to bid at any price level is also related to the bidder's total income and the way in which he is able (or prepared) to 'trade-off' the relative costs of transfer, rent and all the other elements of his total budget (see below). The highest bidder would gain the use of each particular plot of land, so that arrangements of land uses would ultimately reflect the relative rents that each particular user could afford to bid with respect to the costs of transfer and all other elements of production of all other competitors.

Consider, for example, the bid rents which would be made by the respective producers of three different commodities—vines, potatoes and wheat. Naturally, all three would wish to minimise transfer costs and so they bid relatively high for sites in close proximity to the village in which they reside: as distance increases from the village, transfer costs rise and so less is bid for plots at a distance. Assume that the vine farmer can most afford substitution and that he makes the highest bid for sites near the village; the potato farmer might not be able to afford to make such a high bid for a central site but might be able to outbid the wheat farmer for sites at a medium distance away; the wheat farmer would be outbid for all except the furthest sites from the village (figure 7.6A). Given an isotropic surface, therefore, concentric zoning of production would result from the bidding of competing users based on their assessment of the need to substitute rent for transfer and all other costs. There is ample empirical evidence to suggest that such zoning of agricultural production does indeed result from farmers' attempts to minimise movement costs, though the correspondence with theory is only very general, since isotropic surfaces, on which theory is based, do not typify reality (figure 7.7; Chisholm, 1962).

It is not only agricultural production that is characteristically zoned in this way, for the same basic principles apply equally to the location of residential, industrial and commercial activity. Each activity bids against all others for

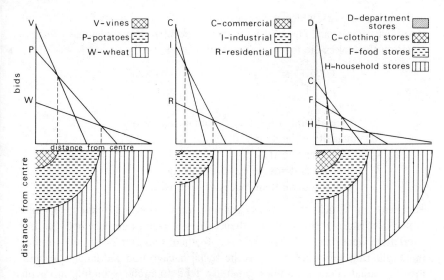

Figure 7.6 BID RENTS A—zoning of agricultural production; B—zoning of urban functions; C—zoning of retail functions within the CBD.

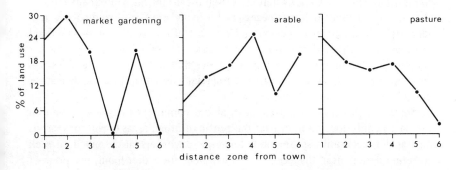

Figure 7.7 LAND USE land use—distance relationships, Abergavenny. Source: Toyne and Newby (1971).

different sites within urban areas and since each site is eventually occupied by the highest bidder, zoning of economic activity (whether residential, industrial or commercial) inevitably results (figure 7.6B; Berry et al., 1959; Isard, 1956).

The bids that any individual householder can afford to make for a residential plot of land are related, like those of the farmer, to his total income which is allocated in different proportions between various sets of costs. The main costs relate to transfer (measured in terms of the cost of commuting to work which varies according to distance involved), and to rent (the cost of the land on which the house stands). The individual's *budget equation* thus takes the form suggested by Alonso (1965)

$$y = P_z z + P_t q + kt$$

where

y = income
P_t = price of land at distance t from urban centre
q = quantity of land needed
k_t = commuting costs to t
P_z = price of all items necessary for living
z = quantity of all items necessary for living

The way in which individuals evaluate these costs is related to their motivation and perception and to their expressions of preference between the alternative opportunities that are open to them (chapter 2). It appears that this evaluation varies according to the social background and income level of the individual. Poorer families generally tend to prefer spending more on household goods than on their houses or commuting, whereas richer families appear to be prepared to spend proportionately more on housing and commuting. It is *proportions* that are so significant in this context since it can also be argued that the 'items necessary for living' are fairly fixed costs which vary little from person to person and that therefore, families with low fixed incomes have little left over in any case for spending on housing or commuting. Given this limitation, it follows that lower-income families will have no other alternative than to bid for relatively cheap housing as near as possible to their place of work.

Since the premises of most industrial and commercial employers were in the past mainly located in or near the centre of urban areas (again for reasons of transfer economy—see below), lower-paid workers also tended to seek accommodation near the centre of towns. On the other hand, the price of land is normally higher near to city centres than it is at some distance away (section 4.2.2; figure 4.16). Low-income workers could therefore only afford a very small amount of land upon which to build, and so areas of working-class housing developed at high density, in small individual units, near the city centre. Richer households could afford greater commuting costs and indeed, usually preferred to get away from the place of work, so they tended to buy up larger amounts of the less valuable land on the outskirts of town, with the result that the peripheral areas were less densely developed. Between the inner and outer zones, housing development was similarly related to class and income considerations, which had their expression in the bid rents offered by different classes of individuals. Using arguments similar to these, Burgess (1925) suggested that a series of generally concentrically arranged zones of housing would emerge around any city (figure 7.8A), and such would indeed be the result if the land surface declined at a directly proportional rate in all directions from a central location. However, in reality, the patterns are more complex than this, because variations in relative accessibility and land values are not so simple as those so far hypothesised.

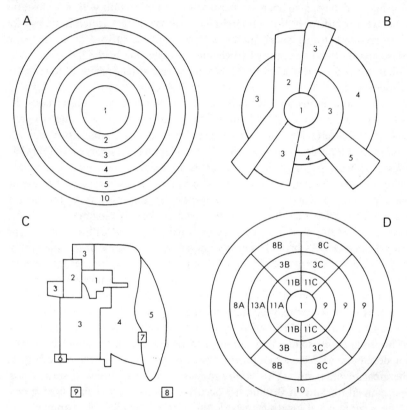

Figure 7.8 URBAN STRUCTURE MODELS A–concentric zone, Source: Burgess (1927); B–sector, Source: Hoyt (1939); C–multiple nucleii, Source: Harris and Ullman (1945); D–a British city model, Source: Mann (1965); *Key* 1, CBD; 2, wholesale/light manufacturing; 3, low-class residential; 4, medium-class residential; 5, high-class residential; 6, heavy manufacturing; 7, outlying business district; 8, residential suburb; 9, industrial suburb; 10, commuter zone; 11, transitional zone; 12, terrace houses; 13, by-law houses (A–middle class, B–low middle class, C–working class).

In the first instance, the existence of major roads leading to the heart of the city tends to make the land in the interstitial areas rather less accessible to the place of work and consequently less attractive from the point of view of housing. Housing development of any given class consequently takes place initially along the major urban roads, and it is only later that infilling of the interstitital areas occurs (Garner, 1960).

Secondly, it is evident that once a particular area is 'invaded' by a particular class of society, a self-perpetuating mechanism may begin to develop. Once a high-class residential zone develops, the value of neighbouring land begins to increase because of its association with a high-status area. As a result, only those householders whose budget function allows for large amounts to be spent on housing are able to afford such sites and consequently the area eventually becomes an extension of the high-class zone. Nor is this process restricted to the buying of vacant land: areas of lower class or older housing peripheral to an existing high-class zone may be bought up by richer households and brought up to their standards. In either case, the result is that the zone of high-class housing successively continues to grow and extend itself. This process of *invasion-succession*, which is essentially similar to the process characterising the development of ecosystems (Hawley, 1950), appears to cause high-class areas to migrate outwards, but within rather limited 'sectors' of the city. The same process may also be true in essence of the shifting location of low-class housing over time, though the sectors into which such housing may extend are heavily circumscribed by the location of higher-class housing which normally tends to 'call the tune' in the housing market. As high-class zones migrate outwards, lower-income groups may invade the vacant premises, converting formerly large houses into smaller units, such as flats. In such circumstances, the process of invasion-succession may cause 'leap-frogging' of development. Through time all the available land in a particular sector may become occupied by housing of a particular kind and may be entirely surrounded by land occupied either by a different class of housing, or by some other non-residential activity, that cannot be bought for development. New development can consequently only take place *beyond* the areas of other uses. This is why so much 'workers' ' housing, such as post war council housing in Britain, has occurred in what may at first sight appear to be a paradoxical location towards the edge of the city. Development thus takes place within sectors and with considerable leap-frogging between different grades of housing (Hoyt, 1939; figure 7.8B).

As the urban area gradually expands outwards it may encroach on other settlements, most of which may themselves have expanded to a lesser extent through the operation of similar processes to those already described. Eventually small villages or even formerly separate towns may coalesce as the processes of urban sprawl continue to create a conurbation or megalopolis whose residential zoning represents the fusion of *several* sectored and zoned developments based on a number of formerly separate nuclei (Harris and Ullmann, 1945; figure 7.8C). It can be appreciated, therefore, that complex structures such as these result from the operation of the mechanisms of movement minimisation and bid rents through successive generations of time.

The industrial and commercial structure of urban areas is based on the same principles. The sites that are thought to be most 'accessible' from the point of view of the industrial or commercial entrepreneur are those at which transfer costs are minimised and for which entrepreneurs are willing to make the highest bids consequent upon an appraisal of their total budget equations. For many activities, and particularly for those that depend on some degree of personal contact with customers, central sites are the most attractive since they represent the points of greatest potential accessibility to the total urban area and its hinterland. Most retailing, wholesaling, administrative, financial and estate business is characteristically located in the very centre of the urban area for this very reason. So also are several manufacturing industries such as tool and instrument manufacturers, printers and newspaper offices, which need to occupy sites of greatest potential accessibility to their labour force and to their markets which are contained within the whole urban area (Martin, 1964; figure 3.5). Within the central area, further zoning occurs as different users bid against each other for the particular sites that they regard as best suited to their needs. Clothing stores, for instance, tend to seek sites at the very centre of the CBD, while supermarkets, car showrooms and industrial or warehousing premises tend to prefer sites on the edge of the CBD where land (of which they need considerable amounts) is less expensive and where there is also greater access and parking facility for the many cars and lorries needing to park or unload outside their premises (figure 3.6).

TABLE 7.3 RETAIL FUNCTIONS AND RIBBON DEVELOPMENT: typical functions of highway-orientated, urban arterial and neighbourhood business streets, all of which are characterised by radial site location in American cities. Source: Berry, Tennant, Garner and Simmons (1963).

Highway Orientated	Urban Arterial
Petrol and service stations	Car repairs and accessories
Drive-in eating places	Bars
Ice-cream parlours	Furniture stores
Motels	Lumber yards/builders yards
Restaurants	Fuel dealers
Fruit and produce stalls	Funeral parlours
	Radio-tv sales and service
Neighbourhood Business Streets	Electrical repairs
Groceries	Discount houses
Meat/fish shops	Florists
Drug stores	Appliances
Liquor stores	Plumbers
Secondhand stores	

For other activities, central locations may not be essential; some retail activities, food manufacturers, financial establishments and engineering and light industries generally, tend to discover that sites adjacent to major routeways into or around the city centre are sufficiently 'accessible' for their purposes, and are also relatively less expensive than central sites. Consequently, they make higher bids than other industrial bidders or private individuals for such sites and the result is that a radial pattern characterises the location of such activities (table 7.3).

7.3.3 *Agglomeration*

Once a particular site is occupied by any activity, there is a tendency for other activities of a similar kind to agglomerate at that site in order to derive the advantages of scale economies (section 3.2). In this way, the location of points and areas in geographical space are closely interrelated since, once the point of movement minimisation (in accordance with Weber's theory) is selected for the location of a particular activity, the establishment of that activity generates its own growth and *zoning* of production (in accordance with von Thunen's theory) inevitably follows.

An important relationship also exists between the principles of movement minimisation and the derivation of scale economies through spatial agglomeration. The producer of any product needs to be as accessible to as large a market as possible in order to maximise his sales; the customer, aware of travel costs increasing with distance, also needs to be as near as possible to the producer if he is to minimise transfer costs. Clearly, both producer and consumer are able to minimise movements by locating as close as possible to each other. Agglomeration may, therefore, arise from the need for transfer costs to be minimised by producers and consumers, and once the process begins, spatial economies of scale will ensure the further derivation of agglomerative locational advantages.

However, as residential and industrial agglomeration proceeds, the relative accessibility of different sites begins to change. The central area in particular becomes increasingly congested: journey times to the centre become longer as more and more traffic brings in customers, workers or materials from increasingly outlying districts, and parking and unloading facilities become inadequate. At the same time, the tendency for factories and retail premises to increase in size in order to derive scale economies, serves additionally to highlight the inadequacies of central locations, since there is rarely sufficient land available there either for new building or for extensions to existing premises (section 3.2). The price of land at the centre tends to rise as more and more entrepreneurs seek an increasingly limited amount of land (section 4.2.2) and this may, paradoxically, cause those same entrepreneurs to reconsider the bids that they are prepared to make for such expensive and congested locations.

In comparison, land at the periphery of the urban area may now appear to possess a new relative accessibility: there is more room to expand, the price of land is lower and, since population has been gradually moving outwards (section 4.1.1), the labour supply may even be in closer proximity than it was to central sites. Similarly, the increasing congestion on radial routes within the urban area normally leads to the building of new peripheral ring roads or by-passes of one kind or another, so that in a physical sense accessibility may well be quite high at peripheral locations.

It is for these reasons that the stage is eventually reached where re-appraisal of the relative accessibility of different sites throughout the urban area may become necessary, and entrepreneurs begin to consider the possibility of substitution between different elements of their total budget equation in order to make revised bids for different sites.

The frictional effect of transfer is thus a fundamental element in the entrepreneur's decision environment, since it underlies the mechanism of movement minimisation, bid rents and agglomeration in the locational system.

FURTHER READING

Blumenfield (1949); Chisholm (1971); Cowan and Fine (1969); Dunn (1954); Greenhut (1956); Grotenwald (1959); Hoover (1948); Kansky (1963); Lean and Goodall (1966); Lever (1972); Pilgrim (1969); Stefaniak (1963); Taafe and Gauthier (1973); Whitehand (1972); Wingo (1961).

8
Demand and Supply

In the locational decision-making process, considerations of demand and supply occupy a position of central importance since together they determine the economic viability of the firm. The volume of demand for a product determines the scale of its production and, through the operation of scale economies, its selling price. Price, however, conditions the volume of demand, so that demand and supply mechanisms operate through the complementary mechanism of price adjustment. This demand–supply–price syndrome both affects and is affected by location; the volume of demand that can be generated at any location is related to variations in the geographical pattern of transfer rates that pertain at that location and, similarly, locational variations in the costs of production (scale, land, labour, capital and transfer) affect the supply conditions of any economic activity (chapters 3–7). Supply and demand are thus related through the determination of price and location which in turn are related to supply and demand conditions in an apparently self-regulating control subsystem.

8.1 Price

The price of a commodity is determined not only by the cost of its production and the level of profit that its maker needs to secure, but also by the level of demand that exists for it at any given moment of time. Gilbert and Sullivan described the condition in a rather more picturesque and lyrical way than most economists, but nevertheless appreciated the basic mechanism

> When every blessed thing you hold
> Is made of silver, or of gold,
> You long for simple pewter.

Demand and Supply

> When you have nothing else to wear
> But cloth of gold and satins rare,
> For cloth of gold you cease to care;
> Up goes the price of shoddy.
> *The Gondoliers*

Through such changes in price, demand and the level of output (supply) are modified, and thereby equilibrium conditions are once more established.

8.1.1 *Demand schedules*

For any commodity there exists at any one time a relationship between its price and demand, that is commonly known as its 'demand schedule'. This schedule normally takes the form of an inverse relationship since the higher the price of a commodity the less of it is demanded and, conversely, the lower the price of a commodity the more of it is demanded (figure 8.1A). The corollary of this principle is equally important: if a large quantity of a particular product is put on the market, it is only able to command a low price, and conversely, if only small quantities of that product are available, it should be possible to command higher prices for it—a principle that occasionally leads farmers to reduce their acreage and industrialists to curtail production in order to obtain better prices for their products.

The size of the relative changes in demand consequent upon price alterations is known as the ELASTICITY of demand and is measured by dividing the percentage change in demand by the percentage change in price.

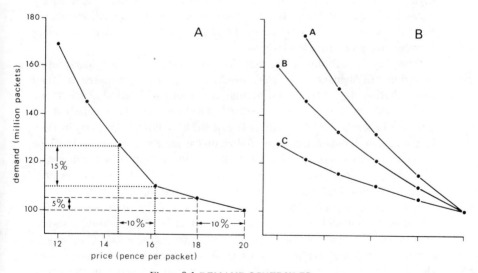

Figure 8.1 DEMAND SCHEDULES.

This ratio is known as the ELASTICITY COEFFICIENT. If a change in price brings about a *more* than proportionate change in demand, demand is said to be ELASTIC, whereas if the resultant change in demand is *less* than the price change, demand is said to be INELASTIC. In terms of the elasticity coefficient, elastic conditions prevail when the value of the coefficient is greater than 1.0, and inelastic conditions prevail when its value is less than 1.0. Should the value of the coefficient be exactly 1.0, the condition is known as UNITARY ELASTICITY. Suppose that a particular model of car sells at £1000 and that at this price the producer is able to sell 20 000 models per week. Because of tax increases and rising costs of production the price may have to be increased to £1100, the result of which is to cause sales to drop to, say, 17 000 models per week. Clearly the increase in price of 10 per cent has led to a drop in sales (demand) of 15 per cent. In this case, the elasticity coefficient is 1.5 (15/10), and demand for the product between these prices is ELASTIC. Alternatively, suppose that a packet of cigarettes sells at 25p and that at this price a retailer sells 1000 packets per week. If the price increases to 28p (a rise of 12 per cent) and demand consequently drops to 900 packets (a decline of 10 per cent), the elasticity coefficient is 0.83 (10/12) and demand is said to be INELASTIC for that commodity between these prices. Patently, the conditions of elasticity for any commodity may vary between different price levels. For instance, a 10 per cent drop in the price of brand X washing powder from 20p to 18p might only generate a 5 per cent increase in demand, whereas a 10 per cent drop in price from 16.2p to 14.6p might produce a larger increase in demand of, say, 15 per cent (figure 8.1A). Under such circumstances, demand for brand X is relatively inelastic at high price levels whereas at lower price levels it is relatively elastic. Elasticity conditions thus refer to the changes in demand that take place between two given price levels *only*.

Elasticity conditions may also be recognised from graphical representations of demand schedules, though great care is required in their interpretation. It is often thought that conditions of demand elasticity are represented by steeply sloping demand schedules, and that inelasticity is reflected in gently sloping schedules (figure 8.1B). While this may very broadly be the case, it does *not necessarily* follow that an increase (or decrease) in slope between any two points on the schedule represents an increase (or decrease) in elasticity, since increasing slope characterises demand schedules that represent constant elasticity conditions. For example, in figure 8.1B, schedule A represents constantly elastic conditions (the elasticity coefficient between all pairs of marked prices being 1.5), schedule B represents constant conditions of unitary elasticity (coefficient 1.0) and schedule C represents constantly inelastic conditions (coefficient 0.5), yet all three schedules are characterised by increasingly steep slopes. It is only by comparing the actual slope between any two points on the schedule and the slope of unitary elasticity between the two price levels concerned that it is possible to recognise whether the schedule between those prices is elastic or inelastic: if the actual slope is steeper than that of unitary elastic conditions, demand between those prices is elastic, and conversely, where it is less steep, demand is relatively inelastic.

The entrepreneur, whether he be industrialist, agriculturalist or retailer, is particularly concerned with the elasticity of demand for his products, because it is this that determines whether the price changes he may have to make from time to time will affect his financial viability. There is, however, an important difference as far as the entrepreneur is concerned between situations of rising prices and falling prices. As we have already seen, a 10 per cent *increase* in the price of a car might, under conditions of elasticity, lead to a *reduction* in demand of 15 per cent. In such a situation, gross revenue is reduced and the firm may end up in financial difficulty. Conversely, however, if elastic conditions of demand prevail and prices are *reduced* by 10 per cent (as in the example of brand X washing powder quoted above), the increased demand yields a larger gross income for the entrepreneur and the firm makes even larger profits. Similarly, there is a marked difference between increasing and reducing prices in situations of demand inelasticity. If prices are reduced in an inelastic demand situation, total revenue falls since the consequent proportional increase in demand is less than the percentage reduction in prices; if, however, prices are increased in such a situation total revenue will increase, since the consequent decrease in demand is proportionately less than the rise in prices.

Conditions of demand elasticity for any one firm may also be affected by the pricing policies of its competitors. If under conditions of elastic demand one firm raises its prices, the conditions of elasticity may be exaggerated since buyers transfer their demand to the alternative supplier whose prices have not increased. On the other hand, the effects of elasticity may be reduced—indeed may even hardly be felt at all—if all producers simultaneously change their prices.

For these reasons, considerations of elasticity may determine the pricing policies of different firms. In turn, pricing policies may have significant geographical and locational consequences. For example, the fact that public transport appears to be operating under conditions of relative inelasticity of demand is one of the main reasons why it is argued that transport rates cannot be reduced (section 7.1.1). If they were, it is doubtful whether the corresponding increase in demand would be more than enough to compensate for the drop in fares, and the industry would only end up further in the red. Because public transport rates cannot be reduced, the public substitute their own private forms for public transport and this, apart from increasing the problem of road congestion, may lead to a further cutting back in the provision of public transport. The consequences of this action may ultimately be to reduce the relative accessibility of the locations from which public transport is withdrawn and this inevitably has important repercussions on the present and future economic and social activity at such locations.

8.1.2 *Supply schedules*

In just the same way that the demand schedule describes the relationship between demand and price, the supply schedule describes the relationship between output levels and price. The supply schedule is similar to the corollary of the demand schedule: normally, as the market price for a commodity increases, more of that commodity is produced and, conversely, as market price decreases, less of the commodity is produced (line A, figure 8.2). Variations in the elasticity of supply occur in much the same way as variations in the elasticity of demand and are similarly reflected in the shape of the graphical supply schedule. If the change in the amount supplied is proportionately greater (either in terms of increase or decrease) than the change in price, supply is said to be ELASTIC; conversely, if the change in supply is proportionately smaller than the change in price, supply is said to be INELASTIC. In fact, it often happens that at certain levels of output, firms are unable to increase supply no matter how much prices may rise, because they simply cannot increase production with existing plant, machinery and techniques. In such circumstances, supply is totally inelastic and the supply schedule is characterised by a horizontal line in the price–output graph (line B, figure 8.2).

As with elasticity of demand, the conditions of supply elasticity may have an important bearing on the profitability of the firm. Likewise, the effect of the supply elasticity for any one product or firm is very closely dependent on

Figure 8.2 SUPPLY SCHEDULES.

the elasticity conditions of other, alternative, products or firms: thus, in the period 1914–34, wheat acreage in Saskatchewan *increased* despite a *fall* in its price, simply because prices had fallen by similar, if not greater, amounts for all the alternative products that could have been grown (Allen, 1954).

The mechanisms of supply elasticity are also very closely related to the costs of production at different levels of output. It has already been shown that unit costs of production vary according to the level of output through the derivation of either economies or diseconomies of scale (chapter 3). Yet output levels are also determined, as has just been shown, by demand, supply and price conditions, so that unit costs of production are also indirectly related to market prices. There is no reason why these two mechanisms should lead to reconcilable output levels: the price mechanism may dictate a production level at which maximum scale economies cannot be achieved and, vice versa, the scale mechanism may dictate a production level at which supply, price and demand are in disequilibrium. The entrepreneur is consequently faced with a problem, but since supply, price and demand *must* be equated (see below), the net result is that the theoretical maximum economies of scale may be unattainable. This then may be a major reason why so many enterprises do not, in reality, operate at a scale that allows maximum economies of production to be derived.

8.1.3 *Equilibrium*

Since price affects both demand and supply, the crucial problem for the entrepreneur in a competitive market is to find the level of production that leads to adequate demand at a feasible price. This can be done by finding the point of equilibrium at which the demand and supply schedules for each product intersect, since the only price that can last is that at which the amounts of product demanded and supplied are equal. If supply exceeds demand (point A, figure 8.3A) prices can only fall, whereas if demand is greater than supply (point B, figure 8.3A) prices are forced upwards. Price equilibrium can only be achieved, therefore, when supply and demand are made to match each other (point X, figure 8.3A).

Supply and demand conditions are constantly altering, mainly due to advances in technology and changes in standards of living. As a result, quite dramatic price changes may be necessary in order to regain conditions of equilibrium. Imagine, for example, that as a result of an innovation in a particular industry, demand for oil suddenly increases. In graphical terms the demand schedule for oil is considerably changed, and a new demand schedule (D′D′), to the right of the original one (DD), is created (figure 8.3B). Oil producers are unable, *momentarily,* to supply this new level of demand, so their supply conditions remain unaltered (the supply schedule thus takes the

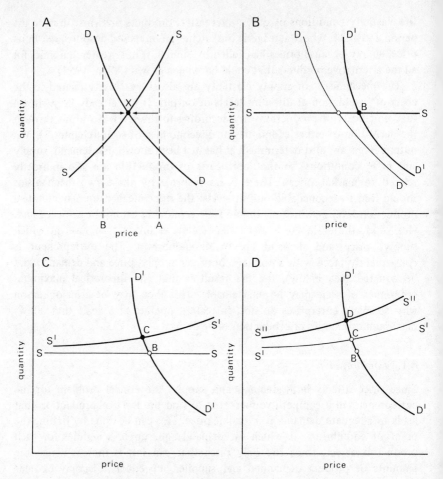

Figure 8.3 CHANGES IN SUPPLY AND DEMAND.

form of a horizontal line [figure 8.3B]). Prices rise, temporarily, to a level that 'rations' demand—the point represented graphically by the intersection point of the supply curve with the new demand curve (point B, figure 8.3B). MOMENTARY EQUILIBRIUM is established, but very soon the oil producers take on more labour and may be able to extract more oil from existing wells until the point is reached where productivity cannot be improved without bringing new oil wells into operation. The net effect of this change in supply is to cause a SHORT-RUN equilibrium to be reached at which the price of oil is less than it was at the point of momentary equilibrium, because the supply curve has been shifted without a shift in the demand curve (point C in figure 8.3C). In the long term, new wells may be brought into use and the supply schedule consequently altered yet again until a new LONG-RUN equilibrium point is reached (point D, figure 8.3D).

In this way, adjustments to price levels and demand follow whenever the supply conditions for any product are changed. The vagaries of nature cause the supply of many agricultural commodities to change from season to season, and it is for this reason that world commodity prices show such marked variations. The effect of a poor wheat harvest, for example, would be to cause a shift in the wheat supply schedule, and with demand unaltered, momentary equilibrium would be found at a high price level; next spring, tempted by the possibility of the prevailing high wheat prices, many farmers may increase their wheat acreage, with the result that the following autumn the supply schedule is altered once more, but this time in the opposite direction. The price of wheat consequently falls quite dramatically to find a new equilibrium.

Similar characteristics typify production in most economic activities, though the precise mechanism by which equilibrium is achieved is dependent on the exact form of both the demand and supply schedule. In this respect, relative elasticity may be a crucial factor in determining both equilibrium price levels and the time span involved.

The significance of these mechanisms lies in their potential effect of creating 'cycles' of production, shortage and surplus in the system. The production of foodstuffs and raw materials for industry is particularly affected, since changes in acreage are the inevitable result of farmers' reactions to changing price levels. Indeed, even the economic prosperity of regions is related to changing conditions of price, demand and supply, for, as Chisholm (1966, p. 56) suggests, 'only the very efficient producers, or those in very favourable locations will be able to make a profit when prices are low, but if prices are high additional enterprises will be set up and established firms will expand their output'.

8.2 LOCATION

Conditions of supply, demand and price all vary spatially, largely as a result of spatial variations in transfer charges, and in just the same way that equilibrium conditions of those factors can be achieved within a firm, so a spatial equilibrium can also be established for different firms and products. In order to analyse the effects of supply, demand and price in creating this spatial equilibrium, certain simplifying assumptions have to be made about all the other factors (such as scale, land, labour, capital and transfer rates) that may affect the location of economic activity. The initial assumption, therefore, in building the theory of spatial supply and demand, is of an *isotropic surface* (see section 2.1.1 for definition).

8.2.1 *Spatial supply*

In a region typified by an isotropic surface, the cost of supplying any location with a particular commodity is dependent upon the distance of that location from the place at which the commodity is produced, since transfer costs are assumed to be proportional to distance (figure 8.4A). Let it also be assumed, for the moment, that the producer varies the price of his commodity to take account of the transfer costs incurred and does not levy an 'average price' irrespective of location. If the producer does not himself distribute his products to other locations, customers have to come to him in order to make a purchase, and in so doing they incur the cost of transport from their home to the supplier and back. In either case the effect is the same: the price varies spatially from the *starting price* (price at the point of production) by an amount equal to the cost of transfer involved. Thus, a retailer at place s may have an article on sale at price P_s but it will cost a customer living x miles from the shop, xt (where t is the transport cost per kilometre) to get to place s, so that the real price to the customer living at that distance is equal to

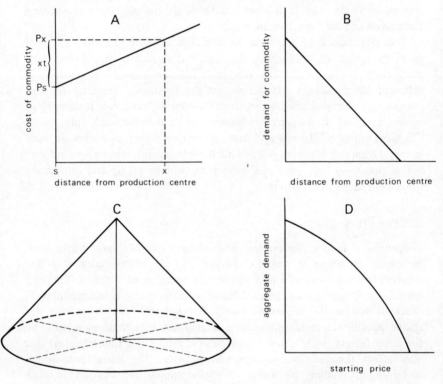

Figure 8.4 SPATIAL SUPPLY AND DEMAND.

$P_s + xt$ (figure 8.4A). In other words, the spatial supply schedule takes the general form

$$P_m = P_s + mt$$

where

P_m = price at distance m from supplier
t = cost of transport per kilometre
m = distance from point of supply
P_s = starting price

For each different starting price (P_s) a different supply schedule exists; the higher the starting price, the less, proportionately, is the additional cost of transfer and, consequently, the 'frictional' effect of distance is reduced.

8.2.2 Spatial demand

It has already been demonstrated that demand for most commodities tends to decrease as the cost of the product increases, though differences in demand elasticity may cause variations on this general theme. Since the cost of the product increases as distance from the supplier increases, demand decreases with increased distance away from the place at which it is produced, until the point is reached where transfer costs make the product so expensive that there is no longer any demand for it (figure 8.4B). In other words, since the quantity demanded d at any location d_m is related to the price at that distance P_m, it follows that demand is a function of starting price P_s, distance m and transport cost t.
That is

$$d_m = P_s + mt$$

Because of the assumption of an isotropic surface, this condition occurs equally in all directions from the point of production so that, geographically, demand for the product is contained within a circular sales area whose maximum extent is determined by the starting price plus the costs of transfer. The total demand D contained within this geographical area can be calculated by measuring the area of the resulting three-dimensional demand cone (figure 8.4C). In fact this is normally done mathematically by integration of the demand function $d_m = P_s + mt$ for all distances out to the edge of the demand cone r and by multiplication of the result by the population density S of the area contained.
That is

$$D = S \int_0^{2\pi} \left[\int_0^r (P_s + mt)\, m dm' \right] d\theta$$

Different sizes of demand cone result for different starting prices and for different commodities, the size being dependent on the extent to which the different commodities are able to bear the cost of transport. Commodities for which starting prices are higher are better able to bear this cost (section 7.1.2), with the result that spatial demand can be derived from a larger tributary area. Higher starting costs, however, entail a reduction in total demand, so that although total demand generally decreases as starting price increases, the geographical area from which that demand is drawn tends to increase as price increases. It is the relationship between starting price and total (or aggregate) spatial demand that is effectively the spatial equivalent of the normal demand schedule since it shows how aggregate spatial demand D varies with changes in starting price P_s (figure 8.4D).

8.2.3 *Spatial equilibrium*

If the aggregate spatial demand schedule is then compared with the ordinary cost curve for any given commodity (note: *not* its spatial-supply curve), the point of equilibrium becomes apparent. In the long run, most producers are able to derive economies from large-scale production with the result that the normal supply-cost curve takes the form shown in figure 8.5A.

The equilibrium point (E in figure 8.5B) represents the point at which demand and supply-cost conditions are balanced and indicates the *starting price P_s* that should prevail. At any other price than this there would be either excess demands or excess supplies of the commodity. Each commodity has a different starting price, and a geographically delimited sales area associated with that price, both of which are determined by the spatial equilibrium of supply, cost and price. In other words, the actual size of the trading area for each commodity is wholly determined by the equilibrium starting price of the product and the extent to which it can bear the cost of transfer. The total demand so created represents the minimum, or *threshold,* demand required for the provision of that commodity.

If these principles applied over the whole isotropic surface, and if all locations were to be supplied with all available commodities, the surface would be divided into a series of circular sales areas, each of different size according to the supply and demand conditions of each commodity. Production would occur at a few 'central places' (so called because the location of the place of production would be central to the sales area), with all non-production centres being 'dependent places' (because they would be dependent on the central place). The central places would, of course, be regularly spaced over the whole surface, at distances determined by their conditions of demand and supply.

However, certain problems arise in trying to fit a series of circles into a

Demand and Supply

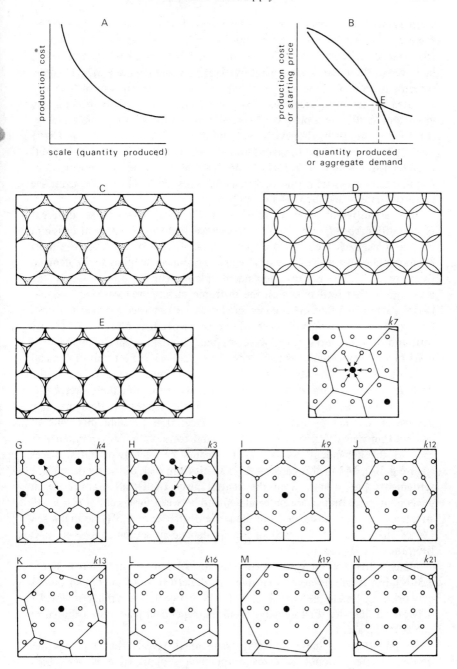

Figure 8.5 SPATIAL EQUILIBRIUM.

given area: if the circles were fitted so that they just touched each other, there would be some areas left where the commodity could not be obtained (the shaded areas in figure 8.5C); on the other hand, if the whole area had to be covered, there would be several overlapping zones (figure 8.5D): for these reasons, neither of these situations represents an efficient way of 'packing' the area. It has been suggested that a more efficient shape for packing any given area is the hexagon (even bees seem to appreciate this simple fact by constructing hexagonal honeycombs!)—though see Haggett (1965, pp. 48–9) for detailed proof. Since hexagons are, in a sense, little more than 'collapsed' circles (figure 8.5E), it is logical and feasible to replace the postulate of circular sales areas with one of hexagonal sales areas. (The same causative factors of supply and demand still apply.)

Just as different sizes of circular trade areas characterise different commodities, so different sizes of hexagonal trade areas represent different commodities, each requiring a different total amount of demand. The most obvious trade area is the one within which a producer supplies all the demand from each of its nearest dependent places (figure 8.5F). Given the assumptions that settlements on the isotropic surface are located in a regular lattice pattern and that the total demand generated by each location is 1 unit, each trading area would, if each of the six dependent places immediately surrounding the central place were supplied, contain a total demand of 7 units (1 unit from each of the six dependent places plus the 1 unit of demand from the central place itself); it is this demand figure that is known as the 'k-value' of the central place. In this particular situation, therefore, $k = 7$. However, not all of the total demand of the nearest settlements may necessarily be met by just one central place: it is possible that, due to competition between producers, the demand for a particular commodity in each of the dependent places may be shared between two central places (figure 8.5G), and in this case $k = 4$ (½ a unit from each of the six dependent settlements plus 1 unit from the central place itself); alternatively, each dependent place may share its total demand between three central places—all situated at the same distance from the settlement—(figure 8.5H), so that $k = 3$ ($\frac{1}{3}$ of the demand from each of the six dependent settlements plus 1 unit from the central place itself).

Many other k-value systems (or 'networks') can be constructed (figure 8.5I-N), but it will be noted that they are all derivatives of the $k3$, $k4$ and $k7$ networks. For example, the $k13$ and $k19$ networks are similar to the $k7$ network in that the total demand of all the dependent places is attributed to one central place.

Examination of these different trade areas reveals that commodities needing a large amount of demand for their support (that is, have high 'thresholds') are provided in fewer central places than commodities with

Demand and Supply 197

lower threshold requirements. It follows that central places offering goods with high thresholds are spaced further apart from each other than those that offer lower order goods. A hierarchy of settlements results, with different combinations of commodities being provided in different places. The precise form of this hierarchy and the location of functions in different settlements is

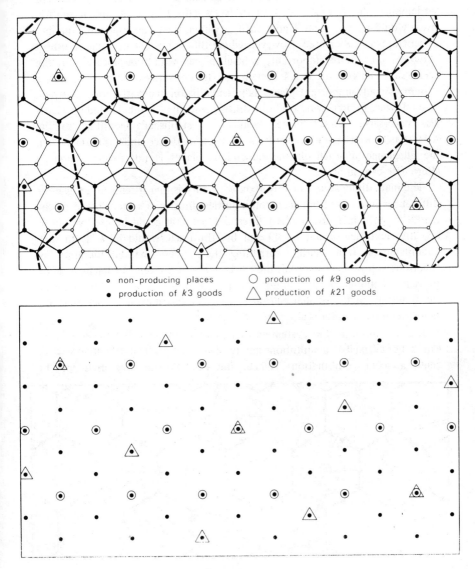

Figure 8.6 CHRISTALLER'S LANDSCAPE.

dependent not only on how many commodities are considered, but also upon the k-networks used to represent them.

The original 'Central Place Theory', suggested by Christaller (1933), was based on the assumption that the k-values in any region would be fixed according to one of three different 'principles'. In areas where the supply of goods from the central places had to be as near as possible to the dependent settlements, the MARKETING PRINCIPLE (the $k3$ system and its derivatives) would be operative. Under such a system, the greatest possible number of central places is created. However, where the cost of constructing transport networks was more important, central places would be located on the TRAFFIC PRINCIPLE (a $k4$ system and its derivatives) because in such a system as many important places as possible lie on one traffic route between larger towns. In areas where administrative control over the dependent places was necessary, central places would be established according to the ADMINISTRATIVE PRINCIPLE (a $k7$ network) since under such a system the allegiance of each of the dependent places is not divided between alternative central places.

The main disadvantage of this 'fixed-k' theory is that it tends to exaggerate differences in demand between commodities; thus, in a system based on the marketing principle, the smallest threshold size that could be used would be $k3$, the next size would be $k9$ and the third size would be $k21$. In such circumstances, it is hardly surprising that a very marked hierarchy of functions is produced, with all the settlements of a given tier having exactly the same combination of functions, and all higher-order places containing all the functions of the smaller central places. The distribution of places of the same hierarchical order is also very regular (figure 8.6).

A rather more flexible system was suggested by Lösch (1940), who relaxed Christaller's 'fixed-k' assumptions and used *all* the possible alternative sizes of trade area in combination. Hence, the $k3$ network was used for the

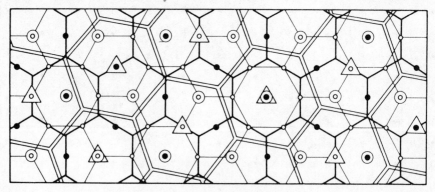

Figure 8.7 LÖSCHIAN LANDSCAPE ($k3$, $k4$ and $k7$ networks).

commodity with the lowest threshold requirements, the $k4$ network for the next largest, then the $k7$, $k9$, $k12$, and so on (figure 8.7). With these 'relaxed-k' assumptions, a hierarchy of functions is still produced, but the combination of functions found in different places is more variable than in the 'fixed-k' system. High-order places do not necessarily include all the same functions as places of lower order, and the distribution of settlements of the same hierarchical order is less regular, though areas containing many functions appear in locational contrast to those containing relatively fewer activities. A system of 'function-rich' and 'function-poor' zones begins to characterise the location of economic activity on the isotropic surface (figure 8.8).

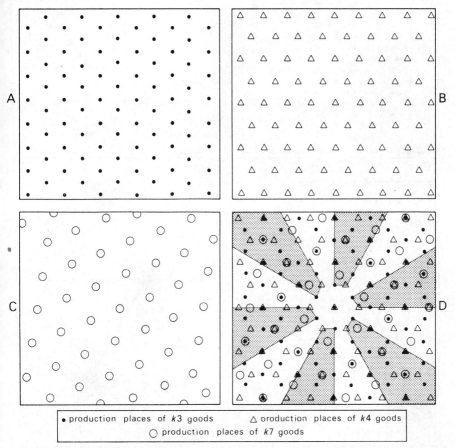

Figure 8.8 LÖSCHIAN LANDSCAPES A–location of $k3$ goods; B–location of $k4$ goods; C–location of $k7$ goods; D–location of all three goods in different spatial combinations showing the development of function-rich and function-poor sectors.

The development of lines of communication between settlements can also be at least partly explained by the operation of demand and supply mechanisms since commodity and human flows should, according to the postulates of central place theory, be in proportion to the size and spacing of central places. Thus Christaller (1933) pointed to the 'strong parallel between the intensity of traffic and the sizes and frequency of central places' and suggested that 'the one can be explained only by the other' (Baskin, 1966, p. 70). He then went on to describe the two alternative network forms which would respectively develop in the central-place systems based on $k3$ and $k4$

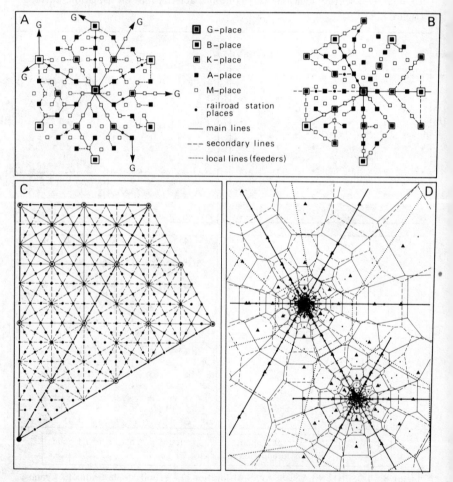

Figure 8.9 TRANSPORT NETWORKS A–Christaller's $k3$ network; B–Christaller's $k4$ network; C–part of the Löschian system; D–system proposed by Isard (accounting for agglomeration).

assumptions (figure 8.9A and B). Lösch (1940) and Isard (1960) (see below), similarly described the communications network that developed in their respective central-place systems (figure 8.9 C and D). The basis of interaction in all these models is thus seen as stemming from demand and supply conditions and, indeed, it is these same conditions that underlie the concepts of *complementarity, intervening opportunity* and *transferability*—the three bases of interaction suggested by Ullman (1957).

In reality, no region will ever follow the precise form of the theoretically generated system of equilibrium because few of the original 'simplifying' assumptions of *ceteris paribus* characterise the real world. If reality is to be more nearly approximated theoretically, the simplified demand-cone schedule must be modified to take account of the factors that have so far either been held constant or even ignored.

The demand schedule hypothesised by Christaller and Lösch was essentially linear in form, since it was based on the assumption that transport costs were proportional to distance (that is $d_m = P_s + mt$). Although this schedule is broadly representative of reality, it is often modified in several ways through the operation of 'tapering', 'blanket' or 'exceptional' transfer rates (section 7.1.1). The spatial-supply schedule may consequently deviate considerably from the straight-line relationship originally hypothesised, with the result that the form of the spatial-demand schedule and its associated cone of aggregate demand is different from that originally hypothesised (figure 8.10).

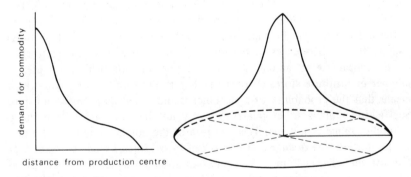

Figure 8.10 MODIFIED DEMAND CONE.

This modification still assumes that customers are always attracted to the *nearest* central place in order to obtain supplies of any commodity. There is ample evidence to show that, in reality, this is far from the normal pattern of customer behaviour (Golledge, Rushton and Clark, 1966). In the Don valley of south Yorkshire, furniture stores are located in each of the three towns of Sheffield, Rotherham and Doncaster; customers from the surrounding villages

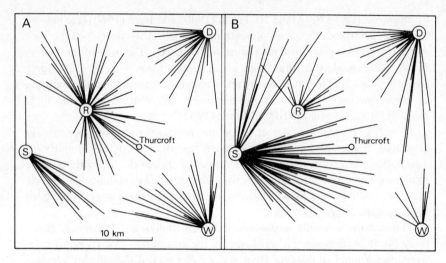

Figure 8.11 DISTANCE AND CONSUMER BEHAVIOUR A—desire lines linking each settlement with the nearest town in which furniture stores are located; B—desire lines linking each settlement with the town which is normally used by local inhabitants (south Yorkshire, 1971). *Key* S, Sheffield, R, Rotherham, D, Doncaster, W, Worksop.

in the region are clearly not attracted to those centres on the basis of distance alone, since in many cases there is a considerable difference between the centre selected and the one that is nearest. Furniture shoppers from the village of Thurcroft, for instance (figure 8.11), tend to visit Sheffield which is 14 kilometres away rather than Rotherham which is only 8 kilometres away.

Several reasons can be suggested to explain this pattern, of which those relating to the differences in the *range* of stores in each of the alternative places appear to be among the more important. Sheffield has a greater number of furniture stores (fifty-eight) than Rotherham (twenty-five), which means that the probability of a customer finding what he needs on a trip to Sheffield is perhaps rather greater than it would be on a trip to Rotherham. Naturally, *range* is not merely a question of the numbers of stores but also the *reputation,* the *quality* and the *cost* of those stores, *as seen in the eyes of the customer.* The customer's perception of geographical alternatives is, as has already been demonstrated (section 2.1.3), rarely very accurate, so that the probability of customers finding what they need in any place is related to the customer's perception of that probability. The customer may also make his choice between the alternative shopping centres on the basis of their relative conditions of parking or congestion. Generally speaking, larger centres have tended in the past to be unattractive in this respect: whether this will continue to be the case in the future, however, remains to be seen, since most cities and large towns are now investing considerable sums of money to improve both their parking facilities and general accessibility.

Various attempts have been made to modify the spatial-demand schedule to take account of these realities. Baumol and Ide (1956), for example, suggest that demand d at any distance m from a central place is related to the customer's expectation of success e, his perception of the costs involved l, his perception of the probability p of finding what he needs in a place offering a number of goods T, the difficulty of shopping at that place, and the costs of giving up alternative activities in order to go shopping c_i. In this way, the original assumption that

$$d_m = P_s + mt$$

can be replaced by the expression

$$d_m = [epT - l(C_n\sqrt{T} + c_i)] - ltm$$

where

$C_n\sqrt{T}$ = the difficulty of shopping measured in costs (C_n) which are taken to be proportional to the square root of size of the centre (with size defined by the number of goods offered T).

t = travel costs per kilometre

From this, it follows that aggregate demand D from a plain of uniform population density S is given by the equation

$$D = S \int_0^{2\pi} \left\{ \int_0^r \left[(epT - l(C_n\sqrt{T} + c_i)) - ltm \right] mdm \right\} d\theta$$

which, when simplified, gives

$$D = 2\pi Sr^2 \left(\frac{epT - l(C_n\sqrt{T} + c_i)}{2} - ltr/3 \right) \qquad 8.1$$

where r = the maximum radius of the demand area (the edge of the demand cone). However, Baumol and Ide (1956) have also demonstrated that

$$r = \frac{epT - l(C_n\sqrt{T} + c_i)}{lt}$$

so that, substituting in equation 8.1 above

$$D = \tfrac{1}{3} lt\pi Sr^3 \qquad 8.2$$

Patently, the assumption of uniform population density also needs revision, for there is ample evidence to show that population density and distance are directly interrelated phenomena, in that density generally decreases as distance from central places increases (see figure 4.6). Furthermore, density lapse rates are far from uniform in all directions around the central city: densities are generally higher and decline less rapidly along the major routeways from city centres than in the interstitial areas and there are considerable regional differences in the form of lapse rates (figure 4.6). In order to account for such variations, Bäumol and Ide suggest that it might be hypothesised that density varies inversely with distance from the central place according to the relationship

$$S_m = S_c/m \qquad 8.3$$

where S_m = density at a point m miles from the central place c
S_c = density at the central place

If such a relationship is assumed, the aggregate demand of the whole demand cone becomes

$$D = S_c \int_0^{2\pi} \int_0^r \left[epT - l(C_n\sqrt{T} + c_i) - ltm \quad dm \quad d \right]$$

which, when simplified, gives

$$D = lt\pi Sr^2 \qquad 8.4$$

In other words, as equations 8.2 and 8.4 show, aggregate demand is related to the maximum distance customers are willing to travel; on a plain of uniform population density, demand varies with the cube of that distance (equation 8.2), whereas on a plain where density is inversely related to distance (in the manner described in equation 8.3), demand varies with the square of the distance (Berry, 1967, p. 85; Cox and Anderson, 1950). Isard (1956) has shown how it may be possible to generate a spatial system of central places based on the assumption that population density does, in fact, decline in proportion to increasing distance; the net result is that hexagonal trade areas are closely packed near to the major nodal points but gradually become larger as distance away from those points increases. This, of course, is as would be expected, since threshold demand can be met from a smaller geographical area when population is densely packed than it can be in areas where population is less dense and more geographically diffuse. It should be noted, however, that some difficulties were experienced in fitting the hexagons under these conditions and, indeed, in some instances it was necessary to replace them with other shapes (Isard, 1956, p. 272).

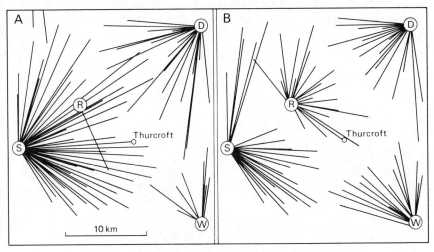

Figure 8.12 STATUS AND CONSUMER BEHAVIOUR A—centres used by shoppers of high socio-economic status; B—centres used by shoppers of low socio-economic status (south Yorkshire, 1971).

Finally, the population in any area has tastes, preferences and habits that are far from uniform, yet it was initially assumed that the population of the isotropic surface was homogeneous in that respect. In fact, because preferences vary considerably from individual to individual and tend to reflect differences in age, sex, income level, status, personality, educational level and mental ability (Huff, 1960; section 2.2), different behavioural patterns are characteristic of different individuals or groups. Murdie (1965), for example, has shown how the shopping centres visited for different goods and services by Canadians from various settlements in southwest Ontario differ quite considerably from those visited by the older 'Memmonites' in the same settlements. Similarly, in south Yorkshire, the central places visited regularly for shopping purposes differ quite markedly between different socio-economic or income groups; shoppers of relatively high socio-economic status in this area tend to use Sheffield rather more regularly than lower status groups who do a larger proportion of their basic shopping in either Rotherham, Doncaster or Worksop (figure 8.12). There is also a significant relationship between status and distance as far as the "attractability' of central places is concerned. As distance increases, the number of shoppers of high socio-economic status drawn to the central place from each settlement increases in proportion to the *total* number of shoppers from those settlements (figure 8.13). The maximum extent of the demand cone may, therefore, be effectively dependent on the attraction of high-status shoppers from peripheral areas. Further modifications to the spatial-demand schedule should thus incorporate some measure of varying tastes, preferences and habits.

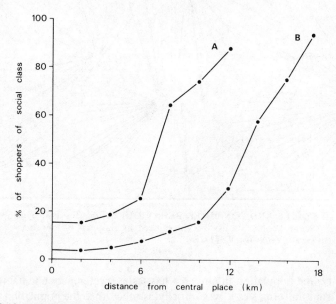

Figure 8.13 DISTANCE, STATUS AND CONSUMER BEHAVIOUR as distance from a shopping centre increases, the percentage of shoppers of high socio-economic status using that centre from each settlement proportionately increases. A–profile around Worksop (north Nottinghamshire); B–profile around Doncaster (south Yorkshire).

Empirical evidence of the location of economic activity tends to validate many of these theoretical conclusions. The basic postulate that there is a minimum 'threshold' demand requirement that varies from commodity to commodity was first verified in an analysis of fifty-two different functions in thirty-three settlements of Snohomish County (USA) (Berry and Garrison, 1958A, 1958B). Having plotted graphically the relationship between the number of establishments of each service type n and the population of the centre in which they were located p, it was possible to establish, for each of the fifty-two services, the average minimum population (threshold size) that was necessary for their support, by calculating from the regression equation of the relationship the value of P where $n = 1$. As had been postulated in central-place theory, it was found that some functions have a larger threshold requirement than others. In this particular region, for example, 196 people were required for the support of one filling station, 380 people for a physician, 729 for a florist and 1214 for an undertaker (Berry and Garrison, 1958A).

Later workers, particularly Haggett and Gunawardena (1964) have tried to find a more refined method of estimating the threshold size of different functions, using a modified Reed–Muench technique to calculate the MEDIAN population threshold. They found, for example, that in southern Ceylon the median threshold population for co-operative stores was 663, while for post offices it was 950, and for dispensaries, 1277.

Many studies have similarly been able to point to the existence of 'hierarchies' of settlements and their functions, from which it is clear that those functions that have low threshold requirements are those that are found in practically every settlement of a region, whereas high threshold functions are located only in larger settlements where their threshold requirements can be met. Many of the earlier studies distinguished between the various settlements of a region on the basis of the functions that they performed, often in a more or less arbitrary way. Bracey (1962) distinguished between first-, second- and third-order villages in central southern England on the grounds that first-order villages had at least twenty shops, second-order villages had ten and third-order villages had five. Similarly, Brush (1953) distinguished between 234 central places in southwestern Wisconsin on the basis of the number of retail and service units they contained, and identified a hierarchy of hamlets, villages and towns. Smailes (1946) used a 'trait complex' based on the presence or absence of banks, Woolworths stores, secondary schools, hospitals, cinemas and weekly newspapers to identify the urban hierarchy of England and Wales. On this basis, the towns of England were graded into five types: major cities, cities, major towns, towns and sub-towns.

Having thus distinguished between the settlements, it was possible to describe what functions appeared to be typical of each level of the hierarchy. In a sense, however, this was rather like putting the cart before the horse, since the basis of the classification was arbitrary. Nevertheless, these early studies were extremely useful in pointing the direction for later work that has attempted to establish rather less subjective bases for the classification of settlements. Berry and Garrison (1958B) applied various tests based on a form of nearest-neighbour analysis, and significance was tested by means of a Chi-squared analysis. By this method, three classes of settlement were identified in Snohomish County (U.S.A.). Later work has shown how cluster analysis and factor analysis may also be used to define the urban hierarchy more objectively on the basis of the functions that are typical of different settlements (Berry, Barnum and Tennant, 1962).

Hierarchies of functions have also been identified *within* urban areas. Berry, Tennant, Garner and Simmons (1963) were the first to identify this intra-urban hierarchy in Chicago. For this city, five levels were recognised

(1) Isolated convenience stores and street corner developments;
(2) Neighbourhood business centres;
(3) Community business centres;
(4) Regional shopping centres;
(5) The Central Business District.

The kinds of shops and services associated with each of these levels have been described elsewhere (Berry et al., 1963), from which it is apparent that the metropolitan intra-urban hierarchy is in many ways rather like the progression from village, through 'town', to 'city' in the general hierarchy of a region.

Work on British cities has revealed the existence of a similar intra-urban hierarchy (Carter, 1972). Smailes and Hartley (1961), for example, point to a three-tiered hierarchy below the CBD in Greater London

(1) Regional centres (for example, Brixton; Kingston)
(2) Suburban centres (for example, Eltham; Wembley);
(3) Minor suburban centres (for example, Shepherds Bush; Hendon);

whereas Thorpe and Rhodes (1966) have identified a fourfold hierarchy, below the level of CBD in the Tyneside conurbation, which is described as comprising centres

(1) Major centres;
(2) Suburban centres;
(3) Small suburban centres;
(4) Neighbourhood centres.

Regularities in the spacing of settlements and their functions have also been recognised in many diverse geographical areas. Indeed, it was precisely because both Christaller and Lösch were impressed by this apparent pattern in reality that they attempted to formulate a theoretical explanation of it. Brush and Bracey (1955) discovered that in both southwest Wisconsin and southern England large towns are about 34 kilometres apart from each other, medium size towns occur at intervals of 12 to 16 kilometres, and small towns are about 6 to 10 kilometres apart.

Thomas (1961), King (1961), Toyne and Newby (1971) and others have pointed to similar 'size–distance' relationships, which tend to support Lösch's

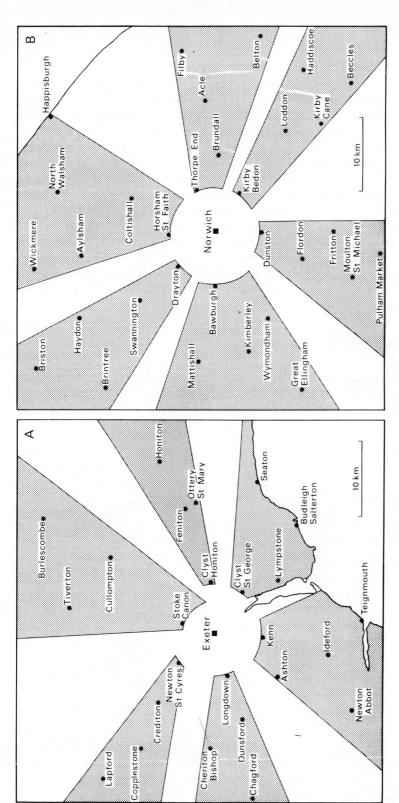

Figure 8.14 SECTORING function-rich sectors (shaded). A—Exeter region; B—Norwich region.

contention that 'all in all, a regular distribution of towns throughout the world is extraordinarily common' (Lösch, 1940, p. 393), except, of course, that the typical distances vary from region to region.

Function-rich and function-poor sectors are also empirically verified. Lösch himself cited the examples of sectoring in the neighbourhoods of Indianapolis and Toledo, and in a general way, sectors or 'corridors' may be similarly recognised round most regional or subregional capitals (figure 8.14).

Empirical evidence thus confirms the general proposition that the spatial organisation and location of functions is fundamentally related to the operation of the factors of demand and supply. Modifications to the spatial-demand schedule that result from relaxation of the original assumptions of *ceteris paribus,* create a theoretical spatial system of central places that more nearly approaches the real-world system, but the changes so wrought are, in every sense of the word, only 'modifications' to the original theory and it is supply and demand that are fundamentally significant in determining the location of economic activity.

FURTHER READING

Carruthers (1967); Davies (1968); Dawson (1973); Marble (1967); Mayfield (1963); Nystuen (1967); Parry-Lewis and Traill (1968); Rushton (1969); Simmons, J. (1964 and 1966); Smith (1970); Stafford (1963); Thorpe and Nader (1967); Ullmann (1968); Ullmann and Dacey (1962).

9
Constraints and Incentives

The locational decision-making process necessitates a careful evaluation and analysis of the relative and combined significance of the conditions of scale, land, labour, capital, transfer and demand and supply. Equally, the decision-maker's motivation, which reflects his beliefs or philosophies of life, may also be significant in determining the decisions that he makes, so that standards of fairness, attitudes, concerns and scruples constitute an essential ingredient of the decision process (section 2.3.1). Such standards vary from individual to individual and, importantly, from individuals to groups. Group standards are normally imposed on the individual by virtue of the individual's membership of the group. The hierarchical structure of firms, companies, industries and of society as a whole is indeed based on the requirement that allegiance to standards or decisions from above should be complete, so that decision-makers at any resolution level of the system may be confronted with a requirement to conform to certain standards that may be imposed either voluntarily or compulsorily by other decision-makers at various resolution levels of the system. From these standards, a series of constraints and incentives are established that may apply to any of the elements of the decision environment either singly or in combination, and that consequently may have a significant effect on the decision-making process.

The fundamental reason for the existence of these constraints and incentives is to be found in the fact that some of the mechanisms of the control system would, if unchecked, create a number of problems of spatial organisation. Positive-feedback relationships in the system have the effect of causing changes to be self-perpetuating until the stage is reached where the structure of the system may be unable to cope with the changing circumstances, and destruction of the system becomes imminent (section 1.2.3). There is some evidence to suggest that certain mechanisms in the landscape control system are of this nature. Cumulative causation is regarded by some observers to constitute one such relationship, and in many ways the net effects of technological innovation appear to constitute another, and to that extent it could be argued that man himself is the ultimate source of all positive feedback in the system. Individuals do not always recognise the potential destruction which they sow and, consequently, constraints on their actions normally have to be imposed by other individuals or, especially, by groups such as governments. The Industrial Revolution, for instance, occasioned a wide range of social and industrial legislation simply because development proceeded in a *laissez-faire* manner in ignorance of the side effects that its technology could produce (Cullingworth, 1970). The problems of overcrowding and congestion, poverty, crime, sanitation, disease and ill-health, which arose in Britain during the nineteenth century largely as a result of urbanisation and 'improved' technology, led to a series of 'Improvement Acts', and especially to the Public Health Act of 1848 which established local Boards of Health to control many aspects of living conditions. In just the same way, it might be argued that present-day planning legislation (of which the aforementioned acts were the precursors) also represents the necessity of legislating against the side effects of modern technology. Certainly, the concern of present-day society with pollution and environmental preservation, which has led to further legislation such as that which controls the heights of chimneys and the permitted levels of discharge of pollutants into the atmosphere, arises almost directly from the introduction of new technological processes. Legislation is not the only mechanism of constraint, however; frequently, financial constraints may be introduced on the decision-maker particularly in the form of taxation.

In other circumstances, the uninhibited interplay of mechanisms in the system may be regarded by different decision-makers as either undesirable (Harvey, 1973) or too slow moving. The introduction of incentives to improve on the 'normal' mechanism, or merely to speed it up, constitutes a further sphere of intervention by the decision-maker himself. Again, such incentives may be of a legal or financial nature and may be introduced at various resolution levels of the system.

A series of feedback relationships, emanating from decision-makers at various resolution levels within the system, thus arise from the introduction of constraints and incentives that may modify the decision-maker's evaluation of the conditions of scale, land, labour, capital, transfer and demand and supply.

9.1 Scale

Considerable economic advantages can be derived by increasing the scale of operation of most activities: unit costs of production may be reduced, under conditions of increasing marginal returns, by the derivation of internal scale economies, and competitors may find locational advantages in agglomerating together spatially because of the possibility of deriving external scale economies. Diminishing marginal returns may, however, lead to considerable diseconomies both internally and externally to the firm, with the result that entrepreneurs in various activities are concerned to establish the optimum scale of operation without incurring any of the disadvantages of being too large or too specialised (chapter 3).

It is principally through the application of technological innovation that large scale operation becomes a practical possibility. It was through the adoption of such innovations as Hargreave's 'spinning jenny' (1764), Arkwright's spinning machine (1769) and Crompton's 'mule' (1779) that the medieval small-scale cotton industry in Britain was transformed into a larger-scale 'modern' industry. Similarly, at the present time, the introduction of automated machines and computers makes possible an increase in the scale of operation of many different firms and industries; while it is mainly through innovations in transport technology, permitting greater and more flexible transfer movements, that the process of agglomeration and deglomeration is facilitated. In these ways, innovation underlies the mechanism of the substitution of capital for land and labour, which in turn is essential for the derivation of scale economies. In order to encourage small producers and small firms to achieve the higher returns that may be achieved from large-scale operation, governments in many nations have legislated for various improvement schemes, and in order to prevent the disadvantages of operation at too large a scale, they have imposed a number of legal (and financial) constraints on many activities and locations.

9.1.1 *Farm reorganisation*

The problem of uneconomic small-scale operation has been particularly evident in the agricultural organisation of many countries, and it is in this sphere that government incentives to reorganisation have been quite extensive.

The operation of many forms of inheritance laws and land-tenure systems tends to create a progressive subdivision of farm units. Under certain arrangements of 'divided inheritance', land is shared between all the sons of the head of the family at the time of the father's death. After a few generations, land holdings inevitably become so small as to be either totally uneconomic or, at best, only marginally productive (figure 9.1). At a time when agricultural productivity can be increased through the substitution of capital in the form of large scale machinery for land and labour, the existence of small-scale farm units is a considerable obstacle to development. Various policies have thus been applied to bring agricultural activity in areas of small

214   Organisation, Location and Behaviour

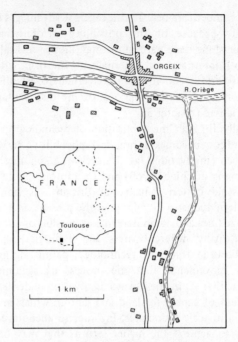

Figure 9.1 LAND SUBDIVIDED THROUGH INHERITANCE location of all the plots belonging to one farmer in the village of Orgeix (Ariège, France).

farm holdings up to a more efficient size standard. In some cases, direct grants have been given to small farms to help them meet their production costs (see section 9.4.4); in others more extensive schemes of farm reorganisation and land consolidation have been implemented. European schemes have been largely based on the French principle of *remembrement*, for it is French agriculture that has had the largest problem in this respect. *Remembrement* began officially in 1941, though prior to that date voluntary exchanges of land between farms had occasionally taken place and resulted in some limited form of consolidation. Some families changed their traditional principles of divided inheritance and agreed to allow all land to be inherited by one heir only, but again such agreements were not extensively adopted by all farming communities. At the beginning of the official policy it was estimated that some 14 million hectares needed consolidation; by 1972 almost 8 million hectares had been reorganised. In addition to this scheme, the Agricultural Orientation Law of 1960 set up Land Development and Rural Establishment Companies (Sociétés d'Aménagement Foncier et d'Etablissement Rural) the main purpose of which is to buy up land, making use if necessary of their right of pre-emption, and then to resell it within a maximum of 5 years; in this way, they are able either to make very small or highly dispersed farms viable or to form new ones. In 1962, further legislation was passed that provides funds to encourage elderly farmers to retire prematurely, so that their land can be reorganised and re-allocated to younger men more conversant with modern farming techniques (Thompson, 1970). By such methods, farms can achieve a sufficient size to enable them to use the modern technical equipment that is necessary if increases in their efficiency and productivity are to be achieved.

9.1.2 *Monopolies, cartels and trusts*

Maximum internal economies can, in certain circumstances, be derived from the amalgamation of individual firms—a process that may lead to the creation of monopolies, cartels and trusts. In England, the first agglomerations occurred in certain specialised divisions of the textile industry (for example, calico printing and coat production), and it was not until the foundation of the Chemical Trust (1926) that a major industrial sector became dominated by a monopolistic organisation. As early as 1919, a Report on Trusts recognised ninety-three monopolistic situations in British industry. Since then, concentration has typified the firms of many British industries. Between 1935 and 1951, for example, quasi-monopolistic situations had developed in a number of major industries such as mining and quarrying, tinplate and refining (Evely and Little, 1960), and as disintegration of the productive process became a more widespread phenomenon, as more and more firms strived to achieve internal economies, the tentacles of quasi-monopolistic financial interests vested in a wide range of firms became successively more far reaching. In 1970, the Secretary for Trade and Industry suggested that 'On the most recent information available, it is believed that at least half the market is in the hands of one company or group for 156 commodities. (The 156 commodities ranging from baked beans to man-made fibres.) In other countries, monopolies first began in the major industrial sectors, as in Germany, where 'the playground of cartels and amalgamations was from the beginning of the movement to be found in the great extractive and heavy industries, such as coal, potash, iron ore, iron and steel and the heavy chemical productions' (Levy, 1966, p. 3). Indeed, in Germany more than anywhere else cartels have traditionally been regarded as a more or less integral part of the normal organisation of industry: in 1905 a Cartel Commission listed 353 quasi-monopolistic associations already in existence, and by 1925 the number had increased to almost 3000. While such monopolistic conditions may be the logical outcome of the operation of scale economies, many governments feel that their existence may not always be to the good of the community that they serve, mainly because they tend to adopt somewhat restrictive policies in terms of output and so can effectively 'dictate' prices, but also because consumers may become entirely dependent on one producer only (Samuelson, 1958, p. 96).

Attitudes, however, vary in detail towards the situation of increasing monopolistic situations. In the United States, constraints on monopolies have been imposed since the implementation of the Sherman Antitrust Act of 1890. German cartel law dates from 1923 and imposes two limitations on cartelisation, namely that any agreements that contain restrictive clauses 'against good morals' are not allowed, and that any agreements that are likely to damage national economic development or that are against the public interest can be equally declared void. In Britain, from 1948, it was possible for the Board of Trade to refer any proposed monopoly in the supply and export of goods to the Monopolies Commission. The Monopolies and Mergers Act of 1965 extended these powers further to include services as well as goods, and if the Monopolies Commission found that a monopoly was against the public interest, it could recommend to the Board of Trade 'much more rigorous remedies than previously, including the prohibition of acquisitions, the forced division of an existing company and the control of prices' (Sutherland, 1969). Since 1948, a large number of enquiries have been made into alleged monopolistic situations in a wide range of industries such as the supply of chemical fertilisers (1959), colour film (1966), clutch mechanism of road vehicles (1968), wallpaper (1964), petrol (1965), cellulosic fibres (1968), contraceptive sheaths (1973), plasterboard (1973) and cross-channel ferry services (1973); and a number of proposed mergers have actually been prevented, for example, the proposed merger of Ross and Amalgamated Fisheries, and that of United Drapery and Burton tailors. But despite the many reports and recommendations, successive governments have often been either slow or reluctant to implement the Commission's proposals (IPC, 1971) with the result that this form of constraint, while potentially significant, has not had its full effect.

9.1.3 Decentralisation

Regional variations in public subsidies within any economic system are apparent in many regions and result from the need to counteract the normal play of market forces and the mechanisms of private investment which, through the operation of external economies of scale, tend to create problems of over-agglomeration, over-specialisation and regional disparities of wealth, prosperity and economic development generally (sections 3.2.2, 6.2).

In France, where these processes have led to sizeable differences in development and prosperity between the Parisian region and elsewhere, one of the main aims of government policies since 1965 has been to slow down the industrial expansion of the agglomerated Paris region and to encourage a

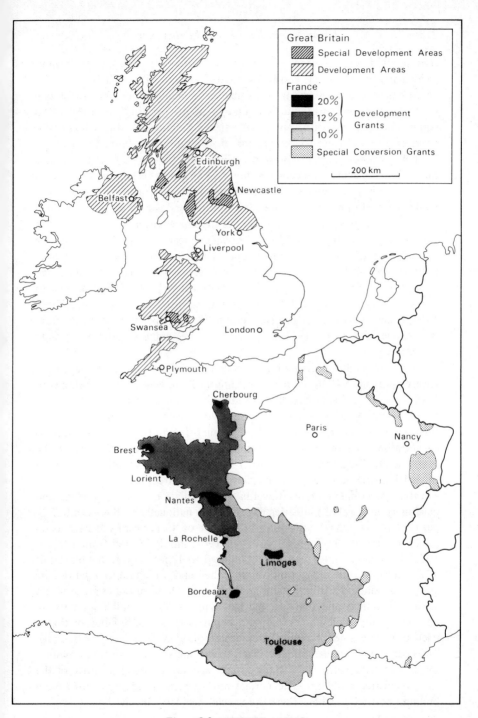

Figure 9.2 ASSISTED AREAS.

substantial decentralisation of industry to other, often stagnating, regions particularly in the west, southwest, and centre of the country. In 1964, development grants were made available for industry moving to certain western areas, and maximum aid (grants of 20 per cent towards capital equipment, together with various subsidies in building and direct loans) was given to firms moving to the regions of Toulouse, Bordeaux, Limoges, La Rochelle, Nantes, Lorient, Brest and Cherbourg (figure 9.2). In addition to these direct financial inducements, the government has established large-scale and long-term programmes to improve the physical infrastructure (especially roads and environment) of those areas to which industry is being constrained to move, and has also set the pace of decentralisation by moving some government-controlled activities to provincial locations: amongst others, the headquarters of the estate-management company of the Armed Forces have been moved to Montpellier, the Navy's hydrographic service to Brest, the Public Dept. Services to Nantes, the National School of Aeronautics to Toulouse and the National Centre of Judicial Studies to Bordeaux. Since 1964, decentralisation has involved about 200 firms per year in moves from Paris alone, though by 1971–2 the rate was beginning to slow down quite noticeably.

In Britain, similar decentralisation policies have been pursued by successive governments since 1948, though it was not until the 1960s that 'Development Areas' as such were extensively introduced. A further category of 'Special Development Areas' was introduced in 1967 (figure 9.2). Various forms of financial (and other) incentives have been offered to encourage firms to decentralise from the London region and to establish new industries in the 'deglomerated' regions. The Industrial Development Act (1966) established a financial distinction between the Development Areas and the rest of the country; government grants towards the installation of certain plant and machinery in selected industries were payable nationally to the extent of 20 per cent of the capital cost, but a higher rate of 45 per cent was payable on assets for use in Development Areas in 1967 and 1968, and from 1969 to 1972 the 'regional rate' was slightly reduced to 40 per cent. In the budget of 1972, however, this distinction was abolished and, for an interim period, the grant was standardised nationally at 40 per cent. Other subsidies are available towards buildings and rents; in the Development Areas, building grants are payable to any eligible undertaking that purchases a *new* building or that is itself providing a new building or extension or adapting an existing building. The cost of building is normally taken to include site preparation, provision for services and permanent fixtures, but excludes the purchase price of the site or existing buildings. The normal rate of grant is 25 per cent of these building costs, but where there are special problems involved in setting up new projects at a 'considerable' distance from a firm's existing undertaking, the rate may be increased to 35 per cent.

Financial help may also be given towards the cost of training labour for particular jobs. Under the Employment and Training Act of 1948, financial assistance may be given towards the cost of training additional labour required by firms moving into or expanding an existing undertaking in a Development Area. The rate of grant varies according to the sex and age of the trainee, with the greatest amounts being paid towards the retraining of males over the age of 18 (£14 per week in 1972). The period for which grant is payable is determined by the Department of the Environment after discussion with the firm, but the period normally allows time 'for a trainee of average ability, with no previous experience in the firm's processes, to acquire the basic skills and knowledge required for the job and to contribute to production without close and continual supervision' (Ministry of Labour, 1967, p. 3). In addition, firms in development areas that are involved in training or retraining of labour as a part of more general measures necessary to prevent a substantial reduction in employment may also be assisted with their training costs. Direct assistance is also provided by means of free training in skilled trades for firms' employees at Government Training Centres. A firm that requires assistance with the training of new workers in semi-skilled engineering work may have the free services of specially trained instructors sponsored by the Government, and there also exists a special Training Within Industry service (TWI) that may be provided free of charge and provides for 'on the job' training of supervisors and operator-instructors involved with assembly lines, processing and packaging work.

Financial inducements are also occasionally offered to selected industries in selected regions through the operation of Regional and Selective Employment Premiums: since such premiums are usually designed to encourage industrial expansion in certain sectors and regions of the national economy, they are usually of temporary duration and may be made to apply for different periods of time from sector to sector or region to region. Thus, for example, in the period beginning 1966, manufacturing industry in Development Areas of Britain was eligible for such a premium, the net benefit of which to the manufacturer was £1.85½ for each adult male employee, with successively lower rates for women, boys, girls and part-time employees. The relative importance of two of the main determinants of location, investment and labour, may thus be manipulated through subsidies in order to encourage developments in certain areas that would otherwise probably not have been chosen. As in France, other incentives mainly in the form of 'example' have been offered by moving some government departments and parts of the Civil Service either to the development regions or to sites peripheral to London: for instance, the National Savings Headquarters have been moved to the Fylde Coast, the Royal Mint has been moved to South Wales, and the Civil

Service Administration has been moved to Basingstoke.

The control of labour mobility at a regional level may also form a further mechanism by which governments may be able to counter the tendency for human activity to agglomerate (section 9.3.2). The operation of these incentives has been such that industrial relocation has been recently biased towards these peripheral areas and away from the over-agglomerated London region, so that earlier patterns of industrial location have been partly counterbalanced (Keeble, 1971).

9.2 LAND

Land is a significant element in the decision environment not only because of the sites that it affords, but also because of the resources that it contains (chapter 4). Certain innovations clearly involve increased demands for larger areas of land. The introduction of large machinery, for instance, of necessity leads to a need for larger sites. On the other hand, most innovations lead to an increased productivity of land since more output can be produced on a given site. This 'substitution' effect between capital and land which is made possible by innovation is clearly apparent, but not all substitution effects are necessarily in the best interests of the population, and it is partly because of this that the control of land and its development in many countries and regions is the subject of extensive legislation or recommendations by government authorities and their agents or advisors.

9.2.1 *Planning law*

Experience of the conditions of life and land use that resulted from unhindered economic and spatial development during the early parts of the Industrial Revolution in Britain led to the recognition of the need to impose certain constraints on development if reasonable social, health and sanitary conditions were to be maintained. The overcrowding and unsanitary conditions of life created by the rapid growth of industry and population during the nineteenth century in certain limited urban areas of Britain were instrumental in causing local governments to enact various by-laws to control street cleaning and lighting and even the layout of certain buildings; and nationally, a series of 'Improvement Acts' and the Public Health Act of 1848 were introduced with the intention of controlling sanitary and health conditions in most of the developing conurbations (Ashworth, 1954).

Further parliamentary acts, beginning with the Housing, Town Planning etc. Act (1909), were brought into force to deal with housing and social conditions generally, but it was not until 1932, when the problems of increasing agglomeration and suburbanisation were first recognised, that planning powers were extended to cover all kinds of land use. None of these acts were binding on local authorities, and there was certainly no attempt to co-ordinate development at a national level. In the 1930s, additional problems of regional imbalance (created through the operation of cumulative causation involved in the development of external scale economies) became so acute in certain areas of the country that a government commission (the Barlow Commission) was set up to study the situation. Because of the onset of the Second World War, the recommendations of this commission were not immediately implemented and, of course, by the end of the war a different problem of redevelopment was beginning to appear, but the Barlow Report was very significant in that it had recommended, for the first time, that the national government should have the power to oblige local authorities to take planning action, and that it should co-ordinate all the local plans to form a broad national planning strategy.

This suggestion became the 'battle cry' of planners towards the end of the war, and was matched by the nation's enthusiasm to build a 'New Jerusalem' from the ruins of the war. The formulation of the 1947 Town and Country Planning Act incorporated the broad suggestions of the Barlow Commission and required all local authorities to prepare comprehensive development plans and to submit them to the newly-formed Ministry of Town and Country Planning for co-ordination and approval, before 1951. Local authorities were to design schemes for the proposed pattern of land development for the whole of their area (figure 9.3), but only twenty-two had done so by the official date and it was for this reason that the system was modified in a new act of 1968, which consequently was embodied in the Town and Country Planning Act (1971). (The latter act merely consolidates the law relating to town and country planning in England and Wales, being the provisions formerly contained in the Town and Country Planning Acts 1962, 1963 and 1968 and certain planning provisions formerly in other acts such as the Civic Amenities Act, 1967. The only amendments to the law effected by this act are minor adjustments recommended by the Law Commission [Cmnd. 4684].)

Local planning authorities (the councils of counties and county boroughs) have two basic functions which are to prepare, modify or revise development plans, and to control development generally. Part II of the 1971 act contains the provisions made in the 1968 act that allow existing 'old style' development plans (that is, those required under the 1947 act) to be replaced by a *structure plan* and other *local plans*. Section 7 of the act relates to the preparation of those structure plans.

7(1) The local planning authority shall ... prepare and submit (to the Secretary of State) a structure plan for their area complying with the provisions of subsection (3) of this Section ...

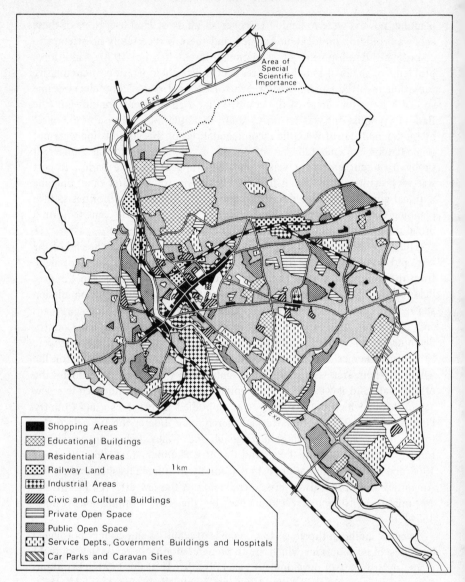

Figure 9.3 DEVELOPMENT PLAN (1947 type). Land-use survey, City of Exeter.

(3) The structure plan for any area shall be a written statement—
  (a) formulating the local planning authority's policy and general proposals in respect of the development and other use of land in that area (including measures for the improvement of the physical environment and the management of traffic);

(b) stating the relationship of those proposals to general proposals for the development and other use of land in neighbouring areas which may be expected to affect that area; and

(c) containing such other matters as may be prescribed or as the Minister may in any particular case direct...

(5) A local planning authority's general proposals under this section with respect to land in their area shall indicate any part of that area (in this Act referred to as an 'action area') which they have selected for the commencement during a prescribed period of comprehensive treatment, in accordance with a local plan prepared for the selected area as a whole, by development, redevelopment or improvement of the whole or part of the area selected, or partly by one and partly by another method, and the nature of the treatment selected.

(6) A structure plan for any area shall contain or be accompanied by such diagrams, illustrations and descriptive matter as the local planning authority think appropriate...

Section 11 describes the preparation of local plans

11(1) A local planning authority who are in course of preparing a structure plan for their area, or have prepared for their area a structure plan which has not been approved or rejected by the Minister, may, if they think it desirable, prepare a local plan for any part of that area.

(2) Where a structure plan for their area has been approved by the Minister, the local planning authority shall as soon as practicable consider, and thereafter keep under review, the desirability of preparing and, if they consider it desirable and they have not already done so, shall prepare a local plan for any part of the area.

(3) A local plan shall consist of a map and a written statement...

(4) Different local plans may be prepared for different purposes for the same part of any area....

(6) Where an area is indicated as an action area in a structure plan which has been approved by the Minister, the local planning authority shall (if they have not already done so), as soon as practicable after the approval of the plan, prepare a local plan for that area.

An essential part of the preparation of such plans is clearly the effecting of a survey of land use which, under section 6(3) is to include the principal physical-economic characteristics of the area of the authority, the size, composition and distribution of the population, and the communications, transport system and traffic of that area. Since development is defined as 'the carrying out of building, engineering or mining operations, in, on, over or under the land, or the making of any material change in the use of any buildings or other land' (TCPA, 1972, section 22), practically *all* forms of land and building development eventually become subject to the control of government. Changes of land use can, however, take place without the developer first seeking planning permission if the proposed use falls within the same general category as the original use, and for this purpose, nineteen 'use classes' were first established in the 1963 act.

Class I. Use as a shop for any purpose except as
 (i) a fried fish shop;
 (ii) a tripe shop;
 (iii) a shop for the sale of pet animals or birds;
 (iv) a cats-meat shop;
 (v) a shop for the sale of motor vehicles.

Class II. Use as an office for any purpose;

Class III. Use as a light industrial building for any purpose;

Class IV. Use as a general industrial building for any purpose;

Classes V–IX. Special industrial uses; (referred to as Special Industrial Groups A–E);

Class X. Use as a wholesale warehouse or repository for any purpose;

Class XI. Use as a boarding or guest house, a residential club, or a hotel providing sleeping accommodation;

Class XII. Use as a residential or boarding school or a residential college;

Class XIII. Use as a building for public worship or religious instruction or for the social or recreational activities of the religious body using the building;

Class XIV. Use as a home or institution providing for the boarding, care and maintenance of children, old people or persons under disability, a convalescent home, a nursing home, a sanatorium or a hospital (other than a hospital, home, hostel or institution included in class XVI);

Class XV. Use (other than residentially) as a health centre, a school treatment centre, a clinic, a crèche, a day nursery or a dispensary, or use as a consulting room or surgery unattached to the residence of the consultant or practitioner;

Class XVI. Use as a hospital, home or institution for persons suffering from mental disorder, or epileptic persons, or a home, hostel or institution in which persons may be detained by order of a court or which is approved by one of Her Majesty's Principal Secretaries of State for persons residing there under a requirement of a probation or supervision order.

Class XVII. Use as an art gallery (other than for business purposes), a museum, a public library or reading room, a public hall, a concert hall, an exhibition hall, a social centre, a community centre or a non-residential club.

Class XVIII. Use as a theatre, a cinema or a music hall.

Class XIX. Use as a dance hall, a skating rink, a swimming bath, a Turkish or other vapour or foam bath or a gymnasium, or for indoor games.

In order to reduce any ambiguities that might arise in the interpretation of the classes, many of the terms used are additionally defined in detail.

'Shop' means a building used for the carrying on of any retail trade or retail business wherein the primary purpose is the selling of goods by retail, and includes a building used for the purposes of a hairdresser, undertaker or ticket agency or for the reception of goods to be washed, cleaned or repaired, or for any other purpose appropriate to a shopping area, but does not include a building used as a funfair, garage, petrol filling station, office, betting office or hotel or premises (other than a restaurant) licensed for the sale of intoxicating liquors for consumption on the premises;

'funfair' includes an amusement arcade or pin-table saloon;

'office' includes a bank, but does not include a post office or betting office;

'betting office' means any building in respect of which there is for the time being in force a betting office licence pursuant to the provisions of the Betting and Gaming Act 1960;

'industrial building' means a building (other than a building in or adjacent to and belonging to a quarry or mine and other than a shop) used for the carrying on of any process for or incidental to any of the following purposes, namely:

(a) the making of any article or of part of any article, or

(b) the altering, repairing, ornamenting, finishing, cleaning, washing, packing or canning, or adapting for sale, or breaking up or demolition of any article, or

(c) without prejudice to the foregoing paragraphs, the getting, dressing or treatment of minerals, being a process carried on in the course of trade or business other than agriculture, and for the purposes of this definition the expression 'article' means an article of any description, including a ship or vessel;

'light industrial building' means an industrial building (not being a special industrial building) in which the processes carried on or the machinery installed are such as could be carried on or installed in any residential area without detriment to the amenity of that area by reason of noise, vibration, smell, fumes, smoke, soot, ash, dust or grit;

'general industrial building' means an industrial building other than a light industrial building or a special industrial building;

'special industrial building' means an industrial building used for one or more of the purposes specified in classes V, VI, VII, VIII and IX referred to in the schedule to this order;

'motor vehicle' means any motor vehicle for the purposes of the Road Traffic Act 1960.

A number of 'conditions' may be attached to any form of development. Normally, permission would be refused for any development that created an obstruction on a road intersection or bend, and also if it required any alteration to be made to the means of access to any classified (that is, major) road. Similarly, certain restrictions on the height, size or elevation of buildings apply to each of the 'use classes'; there are, for instance, limitations on the size, form and location of extensions to private dwelling houses (Cullingworth, 1970, p. 95). Cynics might indeed claim that the Englishman's home is no longer his castle!

Other sections of the act provide for the compulsory purchase of land and for the 'listing' of buildings of special architectural and historical interest. If the authority considers that any use of land should be discontinued or that any buildings or works should be altered or removed in the interests of the 'proper planning' of that area, they may, 'by order require the discontinuance of that use, or impose such conditions as may be specified in the order or the continuance thereof or require such steps as may be so specified to be taken for the alteration or removal of the buildings or works.'—section 51(1). The legislation on developments to listed buildings of special architectural and historical interest, is particularly compelling. Thus, section 55(1): '... if a person executes or causes to be executed any works for the demolition of a listed building or for its alteration or extension in any manner which would affect its character as a building of special architectural or historical interest, and the works are not authorised under this Part of the Act, he shall be guilty of an offence.'

Certain other amenity aspects of land development are also protected by legislation: trees, advertisements and waste land are each the subject of separate sections of the act:

59. It shall be the duty of the Local Planning Authority

(a) to ensure, whenever it is appropriate, that in granting planning permission for any development, adequate provision is made, by the imposition of conditions, for the preservation or planting of trees; ...

60(1) If it appears to a local planning authority that it is expedient in the interests of amenity to make provision for the preservation of trees or woodland in their area, they may for that purpose make an order (referred to as a 'tree-preservation order') ... and, in particular, provision may be made by any such order:

(a) for prohibiting the cutting down, topping, lopping or wilful destruction of trees except with the consent of the local planning authority;

(b) for securing the replanting ... of any part of a woodland area which is felled in the course of forestry operations permitted by or under the order; ...

63(1) Subject to the provisions of this section, provision shall be made by regulations under this Act for restricting or regulating the display of advertisements so far as appears to the Secretary of State to be expedient in the interests of amenity and public safety.

(2) Without prejudice to the generality of subsection (1) of this section, any such regulations may provide

(a) for regulating the dimensions, appearance and position of advertisements which may be displayed, the sites on which advertisements may be displayed, and the manner in which they are to be affixed to the land . . .

65(1) If it appears to a local planning authority that the amenity of any part of their area, or of any adjoining area, is seriously injured by the condition of any garden, vacant site or other open land in their area, then, subject to any directions given by the Secretary of State, the authority may serve on the owner and occupier of the land a notice requiring such steps for abating the injury.

Apart from these kinds of conditions, further constraints may be imposed in order to ensure tighter national control on certain 'key' developments. The requirement that all industrial development in Britain involving an area greater than 5000 square feet must first be granted an Industrial Development Certificate (IDC) by the Department of the Environment, means that there is central control over the national distribution of industrial development. An IDC is made valid for a particular local authority, and the developer must then seek the usual planning permission for a particular site within that area. Office Development Certificates were introduced in 1964 for similar reasons, but applied only to certain regions (London, Birmingham, southern England, west and east Midlands). By such constraints, a two-stage process of regional and local site selection is imposed on development.

9.2.2 Covenants in conveyances

Apart from development controls that are contained in the law of the land, not only in Britain, but in many other countries of the world (Telling, 1967; Laubadère, 1970) some further measures of land control can be effected by the selling or leasing of land subject to covenants as to use. An essential part of most conveyances is the inclusion of such covenants as the following.

(1) Not to erect on the land any building other than one private dwellinghouse with or without suitable garages, such dwellinghouse excluding any garage to contain by admeasurement not less than 800 square feet floor space.

(2) Not to use or permit to be used any dwellinghouse on the land for any purpose other than as a private dwellinghouse and not to use any garages erected on the land or permit the same to be used for any purpose other than as private garages for the use of the occupiers of the dwellinghouse to which it belongs provided that the carrying on in such dwellinghouse of a learned or

artistic profession shall not be deemed to be a breach of this covenant.

(3) Not to erect or permit to be erected on the land any advertisement or hoarding except sale boards or temporary erection or shed of any kind and not to use any part of the land other than that which is built upon for any purpose other than ornamental or vegetable garden and not to do or permit to be done upon the land anything which shall or may be or become a nuisance damage annoyance or disturbance to the Vendor or the owners or occupiers of any land or dwellinghouses in the neighbourhood.

(4) Not to erect make place or build or allow to stand upon the land or any part thereof any hut shed tent booth caravan or boat and not at any time to keep or permit to be kept upon any part of the land any pigs or poultry or to use or permit the same to be used as a breeding establishment for domestic or other animals.

(5) Not at any time to allow any overhead electricity telephone or other wires to be placed over or upon the land unless unable to prevent the same by reason of such wires being so placed under the authority of some statute and in particular not to fix upon any building erected on the land any wireless or television aerials other than those that are confined strictly to the roof or interior of such building.

(6) Not at any time to hang out any laundry or washing on any part of the land so that the same in the opinion of the Vendor spoils the view or otherwise depreciates the amenities of the neighbourhood.

(7) To maintain forever hereafter in good and substantial condition the boundary walls fences or hedges.

However, such private control over land development can only be of very limited effect since the enforcement of the covenants rests entirely and only with the landowner who may simply not have the time or the inclination to inspect whether the conditions are being observed. Nor is there any liability on the part of the vendor or lessor of the land involved to ensure that any of the covenants are 'desirable' in the public interest. For these reasons, the majority of planning law rests with the enforcement of public planning acts such as those already described.

9.2.3 *Standards*

In the process of development planning, local and national authorities usually attempt to establish standards that they consider to be either minimal or desirable. Indeed, the whole concept of development 'zones' (as shown on the development plans) is itself based on the assumption that certain locations are best suited to certain activities and that optimal living and working conditions can be achieved by planning appropriate locations for each different activity. But specific standards in respect of every element of the land environment may also be imposed. Both in Britain and the USA, residential development is encouraged to follow certain recommended density levels: the Parker–Morris standards (table 9.1), while not legally enforceable, are taken as a suggested basis of development in Britain (particularly for housing sponsored by local

TABLE 9.1 PARKER-MORRIS RESIDENTIAL SPACE STANDARDS.

Number of Rooms	Net Floor Space (square feet)		
	1 floor	2 floors	3 floors
1	350	–	–
2	500	–	–
3	650	–	–
4	775	825	–
5	875	925	1050
6	950	1025	1100

authorities); while those of the US Department of Health may serve as a guide to developers in America (table 9.2). Guideline standards have been established in Britain both for residential layouts and for street and footpath

TABLE 9.2 U.S. RESIDENTIAL DENSITY GUIDELINE STANDARDS

Type of Accommodation	Dwelling Units per Acre	
	Desirable	Maximum
1-family detached	5	7
semi-detached	10	12
terraces	16	19
2-storey	25	30
3-storey	40	45
6-storey	65	75
9-storey	75	85
13-storey	85	95

widths (table 9.3). As pressure on land contrives to increase, and as construction technology changes, such standards have to be revised from time to time: the original standards suggested in 1951 were revised in 1965 (Ministry of Housing Circular 27/65). Minimum standards for the provision of recreational, educational and environmental facilities are also being assessed as pressure on the land increases and as modern standards place greater emphasis on the need for open space (table 9.4).

TABLE 9.3 ROAD AND FOOTPATH WIDTHS: Britain, 1965. Source: *Ministry of Housing Circular 27/65.*

Designation of Road or Footpath	Minimum width (ft) of Carriageways	Minimum width (ft) of Footways
Local distributors		
in industrial districts	24	9
in principal business districts	22	9
in residential districts	20	6
Access roads (giving access to land or buildings)		
(a) principal means of access		
in industrial districts	24	6
in principal business districts	22	9
in residential districts	18	6
(b) secondary means of access		
in industrial and principal business districts	20	3 (verges)
in residential districts	13 (or 9 where the street is not >200' in length)	2 (verges)

TABLE 9.4 RECREATIONAL STANDARDS: suggested by US Department of Health.

Type	Approximate Provision	Optimum Site Size
Playground	1 acre/800 people	5–10 acres
Local parks	1 acre/1000 people	2+ acres
Playfield	1 acre/800 people	10–30 acres
National park	1 park/40 000 people	100 acres minimum
Golf course	1 hole/3000 people	18 holes = 150 acres

9.3 Labour

The substitution of capital (in the form of machinery) for labour is perhaps one of the most important ways by which unit costs of production can be brought down *in the long run,* quite dramatically (section 5.3). A major incentive to the industrial entrepreneur who wishes to reduce costs and increase productivity, is thus to substitute capital for labour. Additionally, the entrepreneur is always concerned to keep labour costs, like all other costs of production, at as low a level as possible. In the past, a number of problems relating to labour hours, conditions and wage rates have been created or emphasised by these two characteristics, with the result that over the last century successive generations of governments, industrialists, trade unionists and social workers have been striving to establish conditions of work that would avoid the problems of 'exploitation' or unemployment, yet enable workers to achieve 'a fair day's wage for a fair day's work', in reasonable physical surroundings.

In most countries, therefore, there now exist many agreements between employers and employees, some of which may be legally enforceable, on a wide range of labour issues.

9.3.1 *Conditions of work*

In Britain, the first manifestations of legal constraint on the exploitation of labour are traceable to an act of 1802 which dealt with apprenticing children under the Poor Law. At first, the 'minimum age' constraint applied only to the cotton industry, but by 1833 it had been extended to apply to all forms of textile production. In 1874 the minimum age was raised to 8, in 1901 to 12, in 1944 (by the Education Act), to 15 and in 1972 (through the raising of the school-leaving age) to 16. A similar series of acts and amendments, beginning in 1844, presented legislation dealing with safety, welfare, hours of work and physical surroundings: in 1937 the Factories Act consolidated all the schemes that had been developed by that date, and subsequent extensions to that act in 1948 and 1959 were all consolidated in the Factories Act 1961. Enforcement of this act is in the hands of the Factories Inspectorate, and the legislation relates to general principles of health, safety, welfare, hours of employment and conditions of work. Under the provisions of the act the maximum number of hours' work allowed per week for men and women is 48 plus overtime, but overtime is also limited for women and young persons in that they are not allowed to exceed 100 hours of overtime a year, 6 hours a week and 25 weeks a year, and the total hours' work on any day must not be greater than 10 (10½ if a 5-day week is operated).

Additional constraints in particular industries may be embodied in other laws. Thus the Offices, Shops and Railway Premises Act, 1963, makes provision for the safety, health and welfare of people employed in these premises by enforcing standards of cleanliness, temperature, ventilation, lighting, washing and sanitary arrangements, seating and eating facilities, as well as by setting limits to heavy work or overcrowding. Under section 23 of this act, 'no person may be required, in the course of his work, to lift, carry or move a load so heavy as to be likely to cause him injury', and under section 5(2) employers are obliged to provide 40 square feet (3.68 square metres) of floor space for each person habitually employed to work at one time in the room, and if the ceiling is lower than 10 feet (3.04 metres), 400 cubic feet (1132 cubic decimetres) have to be provided.

Regulations such as these have to be complied with and involve some form of additional expenditure on the part of the employer, which in turn adds to the overall cost structure of the firm or industry. As concern for workers' welfare increases, governments increasingly introduce legislation or establish 'Codes of Practice' to control those aspects of the work environment that are deemed to be undesirable. The recent recognition of problems arising from noise levels in factories has led to legislation in twenty-six countries and to a 'Code of Practice' in Britain that establishes maximum noise tolerations in a wide range of industries.

Other acts have led to the introduction of contracts of employment for all full-time workers in any form of employment (under the Contracts of Employment Act, 1963) and to special arrangements for financial remuneration consequent upon redundancy (under the Redundancy Payments Act, 1965).

The Industrial Relations Act, 1971 adds considerably to the series of parliamentary statutes relating to conditions of work, particularly in that it modifies and extends the written particulars of contracts of employment, and accords new rights to employees who have been unfairly dismissed as well as granting new rights in relation to trade union membership. But the greatest innovation of this act is that it affects the established procedures of the collective bargaining process and establishes a set of forbidden unfair industrial practices which, if committed, allow aggrieved parties to get legal remedies. In one sense, the act is quite revolutionary in that it 'brings changes into areas of life and activity that have hitherto been regulated quite differently' (Pardoe, 1972, p. xvii), but at the same time its basic aims are no different from the aims of many other acts that have preceded it, as section 4 makes clear.

The Provision of this Act shall have effect for the purpose of promoting good industrial relations in accordance with the following general principles, that is to say

(a) the principle of collective bargaining freely conducted on behalf of workers and employers and with due regard to the general interests of the community;

(b) the principle of developing and maintaining orderly procedures in industry for the peaceful and expeditious settlement of disputes by negotiation, conciliation or arbitration, with due regard to the general interests of the community;

(c) the principle of free association of workers in independent trade unions, and of employers in employers' associations, so organised as to be representative, responsible and effective bodies for regulating relations between employers and workers; and

(d) the principle of freedom and security for workers protected by adequate safeguards against unfair industrial practices whether on the part of employers or others.

(Industrial Relations Act 1971, Section 4)

Industrial peace, however, cannot be guaranteed by legislation alone, and the efforts of both management and trade unions to achieve the objectives of growth in output, productivity and real income through voluntary agreements still remains essential if the labour problems of most economies, particularly in the developed nations, are to be solved. Incentives for productivity must be balanced by constraints by both labour and management.

9.3.2 *Mobility*

Government action may also be directed towards the constraint or encouragement of labour mobility. Immigration restrictions and work permits are fundamental and recognised mechanisms of the international control of labour mobility, but national governments may operate equally significant mechanisms to control internal migration, usually in association with other measures aimed at solving the interrelated problems of regional agglomeration and stagnation. In Britain, the Department of the Environment may help to find housing for key workers moving into employment with firms and factories in the development areas, thereby encouraging decentralisation to remote or stagnating regions (sections 5.4 and 9.1.3). Conversely, the Italian government for some time in the post-war years introduced legislation which, by requiring potential migrants to obtain residence permits, was designed to control migration from the relatively poor south to the developing regions of the north.

One of the basic principles of the European Community is that movement between its member countries should be unhampered: that people in one country have the *right* of going to work in any of the others. The possibility that many skilled workers in Britain may be attracted to industrial areas in other member countries of the Community, particularly Germany, constitutes a potential problem arising from such unhindered mobility: 'Yugoslavia has seen its skilled personnel drained away to the fleshpots of Germany, while being stuck with a quarter of a million unskilled and semi-skilled workers. The same could happen here' (Böhning, 1972). Such might be the price which has to be paid for an increase in general prosperity of the whole Community; on the other hand it may be that it will lead to further problems of regional imbalance, which may ultimately necessitate legislative and financial intervention to constrain the mechanism.

9.4 CAPITAL

A major effect of technological development has been to increase the capital requirements of most forms of industry. Most innovations are increasingly indivisible in terms of their capital outlay, and this further increases the level of investment required by modern industry (chapter 6). Together, these two tendencies make it difficult for small firms to take advantage of innovations, since it is only the large firm that is able to make such large-scale investment and that is able to achieve a relatively high rate of capital formation. Similarly, low-level semi-stable economies find capital formation on a large scale through the operation of normal unhampered market forces to be a difficult process, and it is only the already developed economies that are able to generate a high level of savings (section 6.1).

In order, therefore, to offer incentives to firms or to the total economy, as the case may be, governments may find it necessary to adopt a number of different policies, foremost among which is that of taxation.

9.4.1 *Taxation*

By taxation, government takes away part of a country's (or firm's) resources, and either makes them available to private investors in the form of subsidies or loans, or else itself engages in various projects designed to increase both the tangible and intangible assets of the nation. In fact, the two problems of promoting the economic development of the total system while at the same time encouraging investment by individual firms or industries appear, in one sense, to be uncomplementary, for the former may involve increasing taxation while the other might appear to involve reducing taxation or increasing subsidies or loans. This seeming paradox is resolved, however, when it is recognised that taxation tends to go hand in hand with government subsidisation and that therefore all taxation is merely, to adopt a well-known adage, 'robbing Peter to pay Paul and Peter'.

In this way, savings and investment, and consequently economic growth, may be induced through government intervention. On the other hand, taxation has certain adverse side effects: certainly, if the level of taxation is very high, individuals may find that there is little incentive to work beyond a critical minimum level, and governments may consequently find that savings and investment by individuals in the private sector decrease simply because the individual's real net disposable income has been effectively reduced by taxation. In the event, taxation may have the effect of removing the very attitudes that the government was wishing to promote, notably that individual endeavour will lead to increased wealth and a better all-round standard of living. A similar effect can also be experienced by the introduction of various forms of taxation on firms and industries: corporation taxes and other taxes based on trading profits may, if the levy goes beyond a critical point, lead to a reduction in investment by the firms and industries themselves, and ultimately the economy may begin to stagnate as industrial and commercial growth and development recede. It is for these very reasons that governments in the so called 'developed' economies usually pay particularly close attention to the levels of industrial and personal taxation that they impose, and it is also the reason why they may find it necessary to introduce investment rebates and other schemes by which certain capital projects are exempt from taxation (Northey and Leigh, 1971). Despite these problems, it is generally recognised that 'flexible methods of income and profit taxation are a prime requirement for an increased flow of savings' (Mears and Pepelasis, 1961, p. 112), and that they potentially represent a major form of stimulus to economic development.

While it is possible to recognise the potential benefit of taxation there are several difficulties associated with its introduction in low-level semi-stable systems. If incomes are generally very low, it becomes impossible to levy tax at a level that will raise a substantial income for investment; there is the additional problem of deciding who, what and how to tax, and also of persuading individuals that they *must* pay their dues—resistance and evasion, characteristically high in low-level systems, can only be eliminated if the attitudes of all sections of society have been previously changed to recognise the reasons for, and the benefits of, the measures being adopted (section 2.3). Once again, there arises the 'chicken and egg' problem of which comes first. The extent of the income-reducing endogenous processes in the low-level system are apparent: in the early stages of its introduction the tax structure is only able to yield a modest amount—normally, in the experience of most countries, in the region of between 4 and 8 per cent of the total gross national income. Other policies are consequently necessary if investment is to be increased to the 12–15 per cent level that is necessary for the sustained growth of the whole system.

9.4.2 *Inflation*

Inflation has sometimes been suggested as a means by which savings can be forcibly increased, thereby leading to investment generation. The effect of inducing inflation is to cause incomes to lag behind rising prices: it could be argued that manufacturers and industrialists would consider a period of rising prices to be the ideal time to expand their production facilities, and this would effectively necessitate capital investment. On the other hand, it is very difficult to control inflation, as most developed countries have discovered in recent years, and as prices continue to soar the final effect is exactly the reverse of the initial effect—notably that production is curtailed because individuals and firms have less real income. Demand for most products decreases and gross expenditure by the community as a whole is relatively diminished. As demand decreases, workers are laid off, plant is made idle and net *dis*investment begins to occur both by individuals and firms and industries. The induction of inflation is not, therefore, a policy to be generally recommended in trying to generate investment.

9.4.3 *Resource mobilisation*

The only other possibility of generating a sufficiently high level of *internal* savings and capital formation in the total economy is by adopting policies that endeavour to manipulate the resources of the economy, mainly through schemes of labour control. If labour can be mobilised to work on projects that cause an increase either in the total output of the economy or in its tangible capital assets, a sizeable step forward may be possible. In most low-level systems, a contributory factor to the low level of labour productivity is that too many people are unproductively engaged in certain occupations, and that many others are essentially unemployed. If such workers could be encouraged or advised to work on various government-sponsored construction or extraction projects, the effect would be to quicken the pace of development in such capital assets as communication networks, power and raw material sources. Similarly, if the government were to offer various 'rewards' (financial or otherwise) for increased output in every occupation, however menial, the total cumulative effect would be to cause an increase in the total national product. In these ways, labour productivity may be effectively increased and capital assets gradually built up without necessarily having first to find a large amount of money capital. The method, however, like all others depends on a cultural 'revolution': the *ideas* that alternative work forms would help both the individual and the State to achieve greater prosperity, and that increased productivity is essential, must be accepted either voluntarily or, in the view of some more totalitarian governments, forcibly. The complete upheaval that may be necessitated in mobilising labour from one occupation to another and also often from one location to another is sufficient to make many individuals 'opt out', and consequently such schemes may be difficult to operate on a voluntary basis even with great encouragement from governments. It is for this reason that forced-labour mobilisation has not infrequently been chosen as the only way of ensuring development in this sphere.

9.4.4 *Subsidies*

Subsidies are normally made by various government-controlled bodies either to whole industries and activities or to individual firms, and may relate specifically to certain elements of production or generally to the total operation. Whatever form they take, subsidies represent the cost of keeping in business those activities which, although uneconomic, are considered essential for the well-being either of the whole nation or of a particular section of it and, to that extent, they reflect the standards that society considers it necessary to maintain. Without them, locational patterns of production would undoubtedly take a very different form from those prevailing today, since incentives in this form are found to characterise a very wide range of economic activity.

Many forms of subsidy are available for industrial enterprises: loans and grants may be made towards the purchase or building of premises, plant, machinery, equipment and, in some cases, even towards the working capital of the firm. The eligibility of different activities may vary from country to country, from time to time and, within any country, from region to region (section 9.1.3), as changes occur either in the rate of investment generated by the private sector or in the overall state of the national economy. In Britain, for instance, in the late 1960s the government was particularly concerned that private investment in new plant and machinery in certain processes of the manufacturing, extractive and construction sectors of industry was tending to decline to such an extent that future growth might be jeopardised: accordingly, investment grants which extended to a maximum of 20 per cent of the costs incurred were made available to qualifying industries and processes from public funds. (In 1972, the rate was increased to 40 per cent.) In most countries investment grants of these kinds *are* available and, inevitably, their existence has to be considered along with all other elements in the location decision-making process.

Similarly, in the agricultural sector, subsidies from public funds play an important and extensive role not only in determining productivity, efficiency, viability and growth, but also in the entrepreneur's assessment of what to grow, when and where. Again, there are marked variations in the extent to which the governments of different countries give financial support to this sector of the national economy. In Britain, government financial aid to agriculture has traditionally been quite considerable in amount and extent, with guaranteed prices for several commodities being augmented by grants towards many aspects of production. The prices which are 'guaranteed' are reviewed annually so that changing economic and other circumstances can be adequately assessed. In the period from 1966–7 to 1972–3, the cost of this 'Exchequer support' ran at an average level of £125.3m, with prices being guaranteed for the major cereals (wheat, barley, oats and mixed corn), potatoes, fatstock (cattle, sheep and pigs), eggs and wool. Even more was spent by the Exchequer on grants that were made either directly towards production of certain commodities or indirectly towards the cost of farm and land improvement. In Scotland, 'Winter Keep Grants' were introduced in 1964 and were payable to any farmer for growing hay and silage for the winter feeding of sheep and cattle on livestock rearing units. Eligible farms were graded into three classes according to the quality of their land, and a grant of either £2.50, £3.50 or £5.00 per acre was awarded in accordance with their grading. In order to encourage the development of sheep production in the hills and uplands, a Hill and Upland Sheep Subsidy was introduced in 1967, and ever since 1941 a Hill Cattle Subsidy has been available to encourage hill farms to improve the productivity of their hill land and to produce more beef stores for fattening on lowland farms. In 1971, this latter subsidy alone cost £7 023 750. A lowland counterpart of the Hill Cattle Subsidy, the Beef Cow Subsidy, was introduced in 1966 to encourage beef production generally and this cost a further £627 187 in 1971 alone.

Grants for improvement are made for a variety of schemes, ranging from field drainage to farm improvements generally. The Hill Farming and Livestock Rearing Acts (1946–59) provided grants for owners and tenants of livestock-rearing land throughout the United Kingdom who undertook approved comprehensive schemes for rehabilitation of the land. By the end of 1971, these grants, which amounted to 50 per cent of the total costs incurred by the farmer, had been approved for 2812 schemes relating to 5.6 million acres of land in Scotland at a cost of £25 800 000. The Farm Capital Grant Scheme, introduced in 1971, provided for a standard grant of 40 per cent to be made towards capital expenditure incurred on a wide range of works and facilities for the purposes of any agricultural business; field and arterial drainage schemes received a 60 per cent grant, and remodelling schemes necessitated by amalgamation of farms received 60 per cent prior to March 1972, and 50 per cent thereafter. Contributions were paid towards the cost of fertilisers (on the basis of their nitrogen and phosphate content) consequent upon the Agriculture (Fertilisers) Act of 1952; similar contributions towards the cost of liming were also made available, according to the kind and quality of the material and the distance over which it had to be transported, consequent upon the introduction of the Agricultural Limes Schemes of 1966 and 1970.

Other countries have other agricultural support schemes, though most are considerably less generous than those which British farmers have long enjoyed, but all such schemes are normally only regarded as the means by which the agricultural sector of the economy might, in the long run, become so efficient that little or no subsidy would ultimately be required. For Britain, entry into the European Economic Community (EEC) has merely served to encourage the central government 'to adapt the present system of agricultural support to one ... under which the farmer will get his return increasingly from the market' (H.M.S.O., 1972).

Constraints and Incentives   239

The provision of adequate housing, particularly for lower-paid workers is a further sector to which governments may allocate capital funds. In Britain before the turn of the century, all housing was developed privately, and it was really as a result of the housing shortage at the end of the First World War

TABLE 9.5 COST YARDSTICKS: council sponsored housing development, Britain.

Average persons per dwelling	Density in Persons Per Acre. £s—Costs Per Person Accommodated											
	50	60	70	80	90	100	150	160	170	180		
1.0	1993	2073	2203	2313	2898	2467	2584	2584	2584	2584		
1.5	1345	1416	1467	1520	1598	1660	1845	1855	1855	1855		
2.0	1031	1097	1144	1179	1213	1270	1440	1461	1480	1491		
2.5	873	909	953	986	1012	1043	1201	1221	1288	1254		
3.0	773	787	828	860	884	903	1045	1063	1080	1095		
3.5	701	701	740	770	793	812	935	953	968	982		
4.0	648	648	675	704	726	744	853	870	885	898		
4.5	606	606	625	653	674	691	790	806	821	834		
5.0	573	573	585	612	633	649	740	756	770	783		
5.5	546	546	553	579	599	615	699	715	729	741		
6.0	523	523	527	552	571	587	666	681	695	707		

that the central government decided to encourage local authorities to enter into the housing market by guaranteeing to underwrite any loss on the building and letting of such houses beyond that which could be met from the product of an old penny rate. Many changes have subsequently been made to the exact form and extent of this effective subsidy and, indeed, it can be argued that 'the story of the grants subsequently made by the Government towards the provision of housing really amounts to a cataloguing of the circumstances which show increasing recognition that housing is a vital local social service needing government aid' (Hepworth, 1970, p. 164). The Housing Subsidies Act of 1967 went further than merely providing financial subsidy, in that its main aim was to encourage local authorities to provide houses built to the space and heating standards suggested by the 'Parker–Morris Committee' (table 9.1). Subsidies were available for all public housing schemes provided that they conformed to certain cost yardsticks as laid down by the Department of the Environment. These yardsticks were based on the number of persons to be housed in each dwelling, and the overall density of the development (table 9.5). In order to qualify for a subsidy, the costs of development were not to be in excess of 10 per cent of these yardsticks. Thus, the cost yardstick for a scheme of 200 houses, each accommodating four persons at an overall density of 100 people per acre, would be 200 × £744 (the amount allowed per person accommodated) × 4 (the number of persons to be accommodated) = £595 200. If the scheme actually cost more than £601 152 (£595 200 + 10 per cent), subsidy would not be payable on the amount by which £601 152 was exceeded. The precise amount of the subsidy, like many formulae devised by government departments, was arrived at by a complex formula based on the difference between loan charges at current rates of interest incurred on the approved cost of construction, and the charges that would have been incurred had the rate been set at 4 per cent. Had the actual rates been 7.93 per cent (as they were in 1970), the subsidy would have been 0.093p in the £, and the whole scheme would have received a total subsidy of £21 427.20 (595 200 × 0.036p). In addition to this basic housing subsidy, other subsidies were payable for development in special circumstances, such as where subsidence necessitated more expensive forms of construction or where land was particularly expensive. In 1971–2, housing subsidies from the central government represented a grant of 48 per cent of approved costs of all council housing schemes.

Transport subsidies are a further form of government 'incentive' in capital allocation and extend not only to the maintenance of existing networks but also to the development of transport technology for the future (section 9.5).

Quite apart from offering incentives to separate sectors or regions, subsidies may also affect the total economy through the granting of financial aid from one nation to another. Since most aid of this kind takes the form of loans rather than direct gifts, it may represent an effective investment for the lending country. There are normally strict controls over the amount of capital allowed to circulate between nations, in order to prevent currency crises of various kinds, with the result that international capital mobility is considerably constrained and limited. On the other hand, the proportion of a nation's capital formation that may come from countries overseas can be quite high—as was the case, for instance, with many of the member nations of the British Commonwealth in the period 1945–55—and this has repercussions in terms of the pace of development of the benefiting economy (section 6.1.2).

9.5 TRANSFER

Some of the greatest technological innovations over the last century have been in the spheres of transport and transfer, and indeed it is for this reason that it has frequently been suggested that the 'Industrial Revolution' could more properly be called the 'Transport Revolution'.

As improvements in transport technology have altered both the direct and indirect costs of transfer, a general revaluation of the factor of distance and accessibility has occurred, and this has entailed 'a revaluation of almost all of the features, factors and agents that contribute their might to the character of the place' (Mitchell, 1954, p. 228). It is for this reason that changes in transfer technology have had such far-reaching locational effects.

Government legislation and financial support have also contributed to the process of revaluation—legislation may affect the conditions of carriage and employment and subsidies may be necessitated both to induce further advances in transport technology and paradoxically to 'protect' some forms of transport from stagnation, as technology improves.

9.5.1 *Revaluation*

Geographical isolation can only be broken down by the development of intricate and dense communication networks, and it is symptomatic of economic development that communications are developed and accessibility to all locations is increased. Initial 'lines of penetration' are succeeded by the development, first of feeder routes and 'lateral interconnections', and then by 'high-priority linkages' in many underdeveloped areas (Taaffe, Morrill and Gould, 1963). As development proceeds in this general way, nodal structure is initially increased at a local level as most settlements are normally served by carriers of various kinds operating only to the nearest market town (figure 9.4A); nodality and dominance of larger centres thereafter gradually increases as services are extended to cover greater distances, and a more interconnected network results (figure 9.4B).

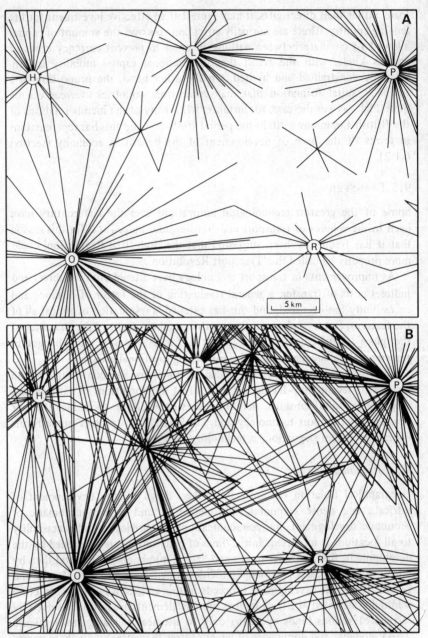

Figure 9.4 LINKAGE AND DEVELOPMENT the development of transport links. A–1886; B–1972 (part of Nord and Pas de Calais Départements, France). *Key* H, Hazebrouck; L, Lillers; O, St. Omer; P. St. Pol; R, Fruges.

Government intervention in the process of network expansion often takes the form of financial incentives and constraints (see below), but the control of network expansion may be more complete in situations where all transport networks are nationalised or state-run. The actual location of transport networks may also be controlled by the inclusion of clauses in the normal planning law of a country that permits local planning authorities to route highways and other forms of communication in accordance with the 'total planning' of the environment. Section 209(1) of the Town and Country Planning Act, 1971, of Great Britain makes provision that 'The Secretary of State may by order authorise the stopping up or diversion of any highway if he is satisfied that it is necessary to do so in order to enable development to be carried out in accordance with planning permission granted under Part III of this Act' (section 9.2.1). Indeed, strategic planning of trunk network systems, including motorway and railway development, is normally either directly controlled, or indirectly supervised, by state agencies at different resolution levels of the system.

Travel *time* has been substantially reduced over the years, not only by the introduction of new means of transport at successive dates, but also by improved technology *within* each form of transport. Railway travel times, for example, have been progressively reduced, particularly recently, and further improvements are inevitable consequent upon the introduction of such innovations as the high-speed diesel train (HSDT) and the advanced passenger train (APT) in Britain and the high-speed turbine trains (TGV) in France. The net effect of such changes is to cause 'shrinkage' of distance (figure 9.5) which in turn may lead to a revaluation of accessibility and comparative location. At the same time, the relative *cost* of distribution has been reduced in just the same way as that of travel time. Increases in travel and freight rates have been at a lower level than those of most consumer items (figure 9.6). Over an 80-year period from 1876, the real cost of transport by ship had fallen by almost 60 per cent and that of passenger transport by air by more than 75 per cent (Chisholm, 1962, pp. 186–7). Such reductions are, of course, made possible by increases in the technical efficiency of each transport form. But costs vary relatively according to the media of transport, and as rates change between the different media, so substitution is likely to occur. In this century, both direct and indirect costs have been relatively most reduced in road transport, with the result that freight and passenger movements have been gradually transferred to this form of transport from both rail and canal. In the last century, it was the comparative lowering of rail rates that led to a similar substitution of rail for canal transport.

One of the biggest cost items of the transport industries' inputs is that of labour, and it is not without significance that labour productivity appears to have been most increased in the transport industries (Chisholm, 1962,

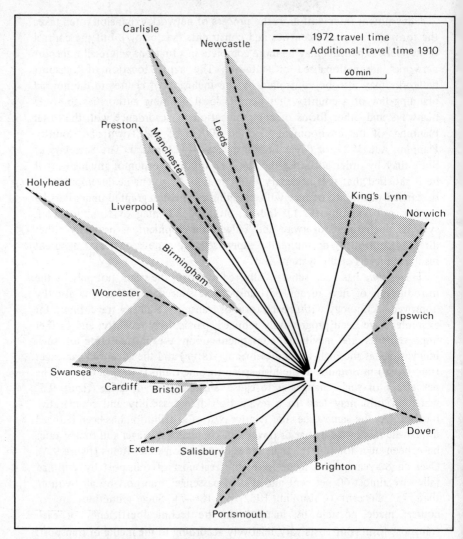

Figure 9.5 SHRINKAGE OF DISTANCE changes in travel times by rail from London to various destinations in England and Wales. Each line is drawn proportional to the journey time between pairs of places: the continuous black line shows 1972 travel times and the additional broken line shows the additional travel time in 1910. The shaded area gives a visual impression of the shrinkage of distance during the period.

p. 186). In this sphere, however, the role of government is normally to act more as a constraint than an incentive, since legislation often 'protects' labour hours and conditions in the transport industries. Safety is, of course, crucial to transport, and is embodied in the common law of many countries. Under

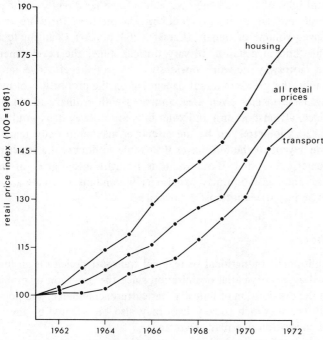

Figure 9.6 RETAIL PRICES AND TRAVEL COSTS.

the contracts of carriage, a private carrier is normally considered to be a bailee of the goods entrusted to him. 'He is, therefore, by virtue of the bailment, under an obligation to apply to the custody and carriage of the goods the utmost care.' (Kahn-Freund, 1965, p. 267). Partly because of this obligation, but also for obvious practical reasons, there is normally some established, and often legally enforceable, code relating to the maximum number of consecutive hours that drivers, particularly, are allowed to work. In Britain, under the Road Transport Act, 1968, drivers of Class A and B (HGV) vehicles are limited to a maximum of 11 hours driving in any period of 24 hours, and to a maximum *consecutive* period of 5½ hours that must be followed by a period of rest of not less than 40 minutes. In combination with the maximum speeds that are either feasible (granted the conditions of the network and the carrier) or allowable (granted that legislation normally prescribes maximum speed limits for HGV), legislation on working hours may have the effect of dictating the effective cost of transport and, consequently, the effective 'range' of transfer by different media. Caesar (1964) suggested that in Britain these circumstances created a 'critical distance' of about 80 miles (129 kilometres): 'up to that distance it is possible for a heavy lorry to deliver its load and return to base within a working day. At about 80 miles, and at multiples of it, costs increase suddenly either directly in the payment of overtime and subsistence, or indirectly in the need for more vehicles to move a given volume of freight' (Caesar, 1964, p. 234). Changing technology causes this 'critical distance' to vary through time: the development of an extensive motorway network obviously increases the effective range quite considerably, the precise amount depending on the proportion of motorway routes available in any given area compared with ordinary routes. Whereas, for instance, a transport firm in Exeter in pre-motorway days would consider Bristol (120 kilometres) to be the effective 'maximum daily reach' to the north, the advent of the M5 makes it possible to double the reach to about 250 kilometres so that Birmingham is brought into range. In this way, innovation and legislation may paradoxically combine to create a constraint as well as an incentive on transfer movement.

9.5.2 *Policies*

The establishment of national or regional transport policies is an important and crucial aspect of spatial organisation since the provision of mobility can influence the distribution of industry, agriculture, commerce, and population; the conditions in which society lives may also be affected by the levels of noise and congestion that may be created.

If standards of service and safety are to be maintained or improved, user charges (either as fares or subsidies) have to be sufficient to cover not only operating costs but also investment costs. The relative charges and standards of different transport media may, in this respect, affect attitudes towards investment: in a situation where a 1 per cent increase in commuting by car can bring chaos to the road system, the social disbenefit of not investing in railways, for instance, may be immense, and further levels of subsidy and investment for railways may thus be justified. Practically everywhere in the world, transport services of all kinds receive subsidies in the form of government grants towards day-to-day costs, often on a quite extensive scale. The reasons for this are many and varied, but centre upon the fact that many services would not be provided if transport operators were dependent only upon the revenue that could be collected from available traffic. On such commercial grounds, services would only be provided where there was commercial justification for the service. Operators faced with relatively high fixed costs (chapter 6) would be most susceptible to this problem and it is for this reason that canal and rail operators have had the greatest problem of economic viability over the last few decades as the distribution of population and the location of economic activity has progressively changed.

However, the provision of a transport service may be justified not merely in commercial terms, but also on social, political or other non-economic grounds. The withdrawal of rail services in rural areas may lead to a decline in social well-being or simply cause hardship to people living in those areas, and on these grounds it might be considered desirable, or even essential, to maintain the service by means of a subsidy from either a local or national organisation or government. Indirect economic gains may also result from the construction or improvement of transport networks, but there may be little direct commercial return to the operators of these networks, who may decide that such construction or improvement was unjustifiable. Yet, as was the case with early railway construction in the USA, 'no line of ordinary importance was ever constructed that did not, from the wealth it created, speedily repay its cost, although it may never have returned a dollar to its share or bond-holders' (Poor, 1869, p. 42). In order to encourage such benefits to the nation, governments may be persuaded, in the national interest, to offer investment grants for the provision of new networks or for the maintenance and remodelling of existing networks. On the other hand, it can be argued that the existence of subsidy to some modes of transport and not to others, results in an uneconomic allocation of traffic so that 'the position which any one of the modes of transport will occupy in the national transportation system will depend in part on the extent to which it or its competitors are subsidised' (Locklin, 1960, p. 827). The cost of subsidies may be so great that even 'social' justifications for their maintenance may appear to be unwarranted, and in this sense the provision of a highly expensive service for a very limited number of transport users could be regarded as a 'disservice' to the community that has to bear its cost. In other words, it could be regarded as irresponsible merely to give subsidies to any transport service in financial difficulty.

Government policies in Britain have clearly been formulated with respect to both sides of the argument. Since nationalisation in 1948, the basic aim of rail and the nationalised sector of road haulage has been to be commercially self-supporting, though certain operations are subsidised where social conditions clearly warrant it. A series of Transport Acts (1947, 1953, 1962) culminating in the Transport Act of 1968 have gradually led to the present attitude that open-ended deficit planning blurs the responsibilities of transport provision. The act of 1968 defined a wide range of operation (all freight operation) in which British Rail must be self-supporting; but it nominated the areas (some passenger services) where financial loss would be carried by the government. Grants in aid of unremunerative rail services are reviewed periodically and are made in some cases for 2-year periods, and in others for 1 year only. In 1971, forty-eight services were granted undertakings for 2 years and seventeen for 1 year only, at a total cost of £16.76m, but total grant aid in the period 1969–71 amounted to £200m and affected the majority of lines in some way or other (figure 9.7). Tapering grants are

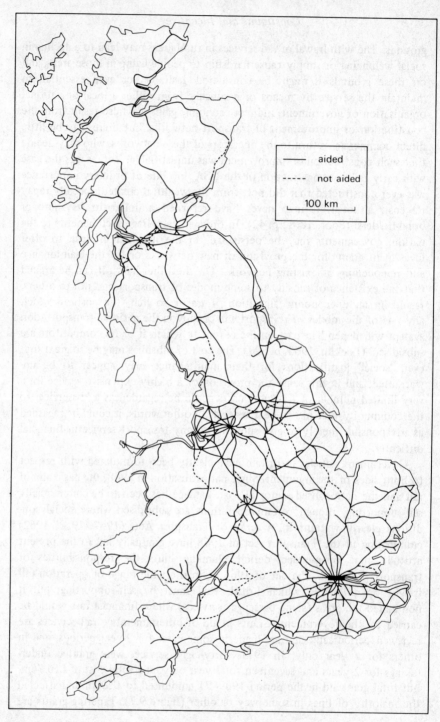

Figure 9.7 GRANT-AIDED RAILWAYS British Rail passenger network, 1972.

also payable to deficiatary Passenger Transport Authorities (PTA) at the rate of 90 per cent in the first year of operation, to be reduced thereafter progressively by 10 per cent in each of the 3 succeeding years, after which the arrangement is to be reviewed: thus the £5m deficit incurred by SELNEC (the authority for southeast Lancashire–northeast Cheshire) in 1971 was supported by a 90 per cent grant from the central government. In Eire, a similar system has been suggested for the Irish Railways (CIE), where the cost of subsidy on the Dublin commuter services would amount to about £400 000 per annum (McKinsey and Co., 1971).

Normally, government aid also extends to the provision of capital equipment and to the development of transport technology generally. The building of motorways (costing £1m for 1 mile) or of airport terminals such as the building of London's third airport at Foulness, are examples of the kind of isolated projects that often receive capital investment grants. The provision of integrated transport networks in which road, rail, air and other services are regarded as complementary is increasingly becoming essential if transfer services are to be efficiently provided over large geographical areas, and governments are beginning to operate policies in many regions that would have that effect. One of the main results of the Transport Act (1968) was the establishment of PTAs that could be set up to provide an *integrated* and efficient system of public transport in certain areas in Britain. Four PTAs were initially established: SELNEC, Merseyside, the west Midlands and Tyneside. Financial aid is available to the PTAs for the major improvement or extension of railway lines, the provision of new rail and bus systems and the construction or improvement of all terminals and interchanges associated with the transport system in their area. Such grants amount to 75 per cent of the approved costs of the schemes: among the first such schemes to be approved were the plans for a new underground rail line for Manchester

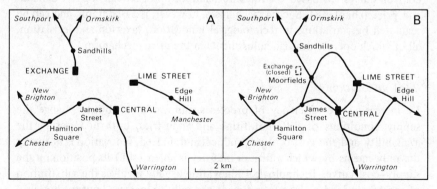

Figure 9.8 RAILWAY DEVELOPMENTS Liverpool area. A–layout in 1971; B–proposed development.

costing £40m and running from Piccadilly via Albert Square and St Peter's Square to Victoria (the Picc-Vic Line), and an inner loop rail line linking the central stations in Liverpool at a cost of £10.8m (figure 9.8).

Grants towards the development of transport technology are often quite extensive and normally take the form of capital grants for projects involving the costly financial development of advanced engines and designs, whether for aircraft (as in the Concorde and Airbus projects) or for railways (as in the development of the HSDT, APT, TGV and similar projects). Investment grants towards the maintenance of standards of safety are also occasionally made. Standards of transport safety are rigidly enforced and have been improved considerably with the application of successive technological innovations. Train safety developments, for instance, have now reached the stage where human errors or mechanical failures have been reduced to an estimated 1 in 100 million chances. But such technology is available only at a cost, and this aspect of transport finance is by no means a small one. The total level of investment required for this and other aspects of train technology in Britain is estimated to be in the order of £125m per year.

Without subsidies, either towards day-to-day running costs or towards improvement schemes requiring large amounts of capital investment, the transport networks of most countries would be a mere shadow of their present aided form, and the effect of this on location and the condition of life would clearly be considerable.

9.6 DEMAND AND SUPPLY

Demand and supply are related to each other through the complementary mechanism of price adjustment and, to a considerable extent, the unhampered play of market forces tends to create an apparently self-regulating control subsystem based on the interrelationships between supply, demand and price (chapter 8). Such a mechanism, however, is based upon, and can be regulated by, variations in technological innovation, taxation and legislation, all of which operate through adjustments in the price mechanism.

9.6.1 *Innovation*

By increasing the efficiency of processes, innovation is able to change the supply conditions of different firms and industries, both in terms of the availability and the cost of their products and, indeed, innovation is normally the only means by which a firm or industry is able to shift its position on the supply–cost curve. Technological innovation usually involves the substitution of capital for land or labour and leads not only to increased output but also, normally, to relatively diminished unit costs (section 3.1), both of which lead to changes in demand, though the extent of the changes depends on the elasticity of supply and demand of the product concerned. Innovation may additionally lead to direct changes in demand for products by making alternative or substitute products available, since alternatives are either more desirable (and consequently people are willing to pay more for them) or less expensive. In either case consumers divert expenditure from one product to another and create a momentary disequilibrium and change in the conditions of supply and demand.

Such changes may eventually create a number of economic and social problems in any region or economy, since demand changes may lead to the eclipse or decline of older industries and give rise to considerable problems of unemployment or regional stagnation. The growth of the artificial-fibre industries is a case in point. In less than three-quarters of a century since the first commercially successful factory for making an artificial fibre was set up in Besançon by Count Hilaire de Chardonnet, the proportion of world fibre production accounted for by synthetic fibres has increased to 30 per cent. In the wake of this development, areas whose prosperity was based on, or associated with, traditional fibre production, such as Lancashire, have suffered a sizeable economic decline. Technological innovation may, in this way, lead to disequilibrium in the demand and supply mechanism, and to problems of a geographical and social nature.

9.6.2 *Taxation and tariffs*

The effect of the imposition of taxation of one kind or another is similar to that of technological innovation. As taxation increases, either on individual products or on gross personal incomes, demand correspondingly decreases, though the extent of the influence on any commodity is dependent on its elasticity. Likewise, costs of supply increase if taxation increases. The effect of taxation is thus to cause a shift in the supply curve, or a movement along the demand curve (*not* a shift in the latter since it is merely price that changes, not demand at *all* prices [figure 9.9]).

A special form of taxation is to be found in the practice imposing tariffs on certain commodities in trading: through the imposition of tariffs or through the introduction of certain commodity agreements, constraints are imposed on the incentives of specialisation and free trade.

It is generally argued that in any economic system 'unhampered trade promotes a mutually profitable international division of labour, greatly enhances the potential real national product of all countries and makes possible higher standards of living all over the globe' (Samuelson, 1958, p. 672). Just as specialisation can increase the productivity of the firm (chapter 3), so the productivity of international (or interregional) economies

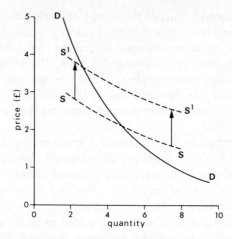

Figure 9.9 TAXATION AND THE SUPPLY CURVE the curve S^1-S^1 represents the effect of a £1 tax levy on a commodity at all quantities of production.

can be increased by each economy specialising in the production of those commodities for which it has a 'comparative advantage' over its competitors. But in order to specialise in this way, it is necessary that there be unhampered movement of goods, people and capital between all regions in the trading system.

The essential truth of this proposition can be verified by considering two regions, north and south, which are both physically able to produce two commodities, wheat and butter. Suppose, initially, that the north is able to produce, with a given amount of labour, either 50 units of wheat or 25 units of butter, and that the south can produce either 25 units of wheat or 50 units of butter. In other words, for every 50 units of wheat production in the north, 25 units of butter production are sacrificed (that is, for every 1 unit of wheat produced, ½ unit of butter is lost), whereas for every 25 units of wheat produced in the south, 50 units of butter are sacrificed (that is, for every 1 unit of wheat produced, 2 units of butter are lost). The north thus has a comparative advantage over the south in terms of wheat production. By similar reasoning it can be seen that the south has a comparative advantage in terms of butter production: for every 50 units of butter it produces, it only loses 25 units of wheat (½ unit of wheat for every unit of butter), whereas for every 25 units of butter produced in the north, 50 units of wheat production are lost (that is, 2 units of wheat for every unit of butter). If the north were to specialise in wheat production, both regions would be able to benefit, provided trade could take place freely. As far as the north is concerned, the only limiting factor to trade is that it must be able to get *more* than ½ a unit of butter in trade for every unit of wheat it exports, and this it will patently be able to do, since the south would be willing to give up to 1 unit of butter for every ½ unit of wheat they could get in trade. Similarly, the south would need to get *at least* a ½ unit of wheat for every unit of butter it exported, and, as we have just seen, the north would be willing to trade up to 1 unit of wheat for every ½ unit of butter they could get in trade. Trade is thus to the advantage of both sides, given the existence of this kind of comparative advantage.

In this particular example, of course, it should be no surprise that trade would occur since the north had, in any case, an *absolute* advantage over the south in terms of wheat, and the south had an equally absolute advantage over the north in terms of butter (that is, they could each, respectively, produce a greater total quantity of one commodity than the other). But even when one of the regions has an absolute disadvantage, it may be that it still has a comparative advantage, which would enable specialisation and trade to occur. Suppose that the east region could produce 40 units of wheat or 30 units of butter, and that the west could only produce either 20 units of wheat or 10 of butter. The east clearly has an absolute advantage for the production of both commodities, and yet there is still a comparative advantage to be found in such a situation. For every 40 units of wheat produced in the east, 30 units of butter production would be sacrificed (that is, for every 1 unit of wheat produced, ¾ units of butter is lost), whereas for every 20 units of wheat produced in the west, 10 units of butter would be sacrificed (that is, for every 1 unit of wheat produced, ½ unit of butter is lost). In this situation the west has a comparative advantage over the east in terms of wheat production. By similar deduction, the east has a comparative advantage in butter production, and specialisation and trade can be seen to be both desirable and feasible.

By extension, this argument is applicable to the production of several commodities in many regions, and helps to explain why there is an incentive to be found in trading and specialisation that may lead to the production of commodities in locations which, in absolute terms, may appear to be the least well-endowed of all the available alternatives. It follows, as Chisholm (1966, p. 43) has suggested, that 'the real explanation for any particular pattern of localisation, say cotton manufacture in Lancashire, may lie as much in the relationships of cotton to other types of enterprise in Lancashire and elsewhere as to relationships directly between cotton textiles and the factors of their production'.

It is, however, rare that unhampered trade is allowed to take place between nations. Through the imposition of tariffs or through the existence of certain agreements, constraints are imposed upon the incentives to specialisation and free trade. The effect of a tariff duty on imports into any country is to raise the price of imports to the consumer who, it is assumed, will then find the price of his own country's goods comparatively less expensive than an equivalent imported commodity. In this way, the terms of trade of the trading nations are modified through adjustments to their comparative advantages, and domestic producers, particularly those who are relatively inefficient, are 'protected' from foreign competition. In some instances, the real justification for tariff protection may be based on non-economic grounds: it may be thought desirable not to be entirely dependent on foreign countries for certain vital commodities, for reasons of 'the national interest' or for reasons of defence; alternatively, it may be that social problems of unemployment, redeployment or migration may arise if certain 'key' sectors of the economy were not 'protected' from external competition. On the other hand, tariffs may be justified on economic grounds, in that total specialisation leaves any economy open to price fluctuations over which it may have little control, and serious balance-of-payments problems may develop as a result. Similarly, it is often argued that tariffs are essential for the protection of 'infant economies'. Developing nations, for example, may feel that tariff protection is the only way by which free trade can be constrained and their economy allowed to develop to the stage at which it is able to derive reasonable terms of trade (Johnson, 1967).

Most nations have experienced phases of 'protectionism' behind tariff barriers, alternating with periods of movement towards freer trade. Britain's first attempts towards freer trade with Europe began with the lowering of import duties from France in 1784, and for the next 100 years a series of treaties and agreements successively lowered tariffs on a wide range of industrial and agricultural commodities. The Cobden–Chavalier treaty of 1869, for instance, pledged Britain and France to a reciprocal reduction of tariffs, and prior to that, in 1845, great controversy had been sparked off in Britain when Peel suspended the Corn Laws and, a year later, asked for their permanent repeal in order to encourage free trade. Similar attempts were characteristic of many European countries at the same time: the Prussians organised a customs union, the Zollverein, to allow free trade with Germany, and the French in turn agreed similar treaties with most other nations. Practically all these treaties contained the famous 'most-favoured nation' clause, the effect of which is that each signatory grants to the other signatory all the concessions that it has previously given to other countries. Thus, when France signed with the Zollverein, it gave the concessions that it had previously granted to England in the Cobden–Chavalier treaty.

From the 1890s, however, the free-trade movement was eclipsed by a long period of protectionism that resulted very largely from increased competition from both the USA and Russia. Prices for many products began to fall, and consequently many nations decided that they must protect their own interests and establish tariff barriers round most of their products. The world economic depression of the 1930s prompted even greater protectionism and many special commodity control schemes were agreed between nations. The output of tin, tea, rubber, sugar, wheat and copper became internationally controlled by a system of production quotas under which exporting countries agreed to accept a schedule of export quotas and not to increase their own domestic production (Kenen, 1964).

In order to promote as rapid a recovery of world trade as possible after the Second World War, several attempts were made to establish multilateral trading agreements. An International Trade Organisation resulted from conferences held under the auspices of the United Nations in 1947-8, and a General Agreement on Tariffs and Trade (GATT) was internationally established. Tariff reductions followed, often on a quite extensive scale (table 9.6), and certain international commodity agreements (ICA) were also

TABLE 9.6 TARIFF REDUCTIONS: USA 1947-62.

Commodity	Tariff Levied (%)	
	1947	1962
Agricultural products	11.7	8.6
Chemicals	19.6	12.3
Earthenware	35.4	22.8
Metals	18.8	11.7
Sugar	13.5	9.0
Synthetic fibres	31.0	16.8
Tobacco	34.7	20.8
Wood	7.0	3.5
All durable imports	16.5	12.0

deemed to be acceptable under certain circumstances. Any member of GATT can ask the United Nations to set up a study group with a view to establishing an ICA if it feels that either widespread unemployment or considerable surpluses cannot be avoided by the normal market forces under free trade conditions. Four schemes, for wheat, sugar, tin and coffee have so far been agreed in this manner.

At the same time, the move towards freer trade has been manifest in attempts to establish various customs unions or trading blocks within which all tariffs are removed and replaced by some form of common external tariff. In Europe, the Benelux union of Belgium, the Netherlands and Luxembourg, inaugurated in 1948, led once more to the recognition of the advantages to be derived from free trade, and the Schumann Plan led to the signing in 1951 of an agreement which established the European Steel and Coal Community

(ECSC) incorporating Benelux, France, West Germany and Italy. Seven years later, these same six nations extended the steel and coal agreement to cover all commodities, and the European Economic Community (EEC) came into being with the intention of eliminating all internal trade barriers, establishing a common external tariff and a common policy for agriculture, industry and transport, as well as forming a political union, all over a 12-year period extending from 1958 to 1969. In similar vein, the grouping together of seven countries, Britain, Austria, Denmark, Norway, Portugal, Sweden and Switzerland, in 1960, to form the European Free Trade Association (EFTA) was designed to allow reductions in tariffs, by stages, between the participant nations. The enlargement of the EEC in 1973 consequent upon the applications of Britain, Ireland, Norway and Denmark to join that community, represents the extent to which the economic advantages of free trade are currently recognised. Indeed, the existence of many such free trade areas in the world (such as LAFTA, CACM, COMECON) is surely testimony to the economic incentives of free trade in comparison with the constraints that tariffs necessarily impose.

FURTHER READING

Cullingworth (1973); Freeman (1968); Goodall (1970); Harvey (1973); Linge (1967); Maudsley and Burn (1970).

10
Interdependence

The fundamental characteristic of systems is that of interdependence, since it is the existence of the various direct, looped, series and parallel links between the elements and their attributes that gives rise to system structure and performance, both of which are themselves mutually interdependent.

The *morphological* structure of human landscape systems is based on the interdependence of the attributes of their residential, industrial, commercial, agricultural, transportational and recreational elements at various resolution levels (figure 1.7). The residential element of the landscape system is thus related directly to the industrial, commercial, agricultural, recreational and transportational elements in terms of its respective and relative location, age and conditions, but it is also related indirectly to each of those elements: for example, a relocation of residential property may cause a relocation of commercial activity that in turn may cause a revaluation of the location of all other activities. Similar interdependent causal relationships characterise all the other elements of the landscape system, so that a change in any of the attributes of any of the elements, either separately or in combination, may have repercussions varying in proportion to the strength of the interdependence between each pair of elements.

Changes in the attributes of the elements of morphological structure result from the various inputs and outputs emanating from the control system—an equally complex structure based on locational and environmental decision-making processes. The decision-maker's evaluation of each of the elements of the decision environment is dependent on his personal characteristics, information, preferences and motivation, all of which vary from individual to individual and between individuals and groups at various resolution levels (chapter 2), and in this way the decision-maker himself constitutes a further subsystem within the control system (figure 2.11). The elements of scale, land, labour, capital transfer, demand and supply and the various constraints and incentives that have been introduced by successive decision-makers, together comprise the decision environment, and the decision-maker's assessment of the significance of any one of those elements is inevitably dependent on his assessment of each of the other elements. Scale economies, for instance, affect the conditions of land, labour, capital, transfer, demand and supply and at the same time are affected by various constraints and incentives in the form of legislation or financial aid (chapter 3); similar interdependence characterises each of the other elements of the decision environment (chapters 4–9).

The landscape control system is based on an intricate mechanism of interdependence that operates through a myriad of direct, looped, series and parallel relationships between decision-makers at various resolution levels and each of the elements of their decision environment (figure 1.7). It is this control system that is responsible for the *performance* of the landscape system, since the inputs and outputs that take place between the elements and attributes of the landscape, result from the various decisions made by different individuals and groups about the organisation and location of residential, industrial, commercial, agricultural, transportational and recreational activity. Again, interdependence characterises the effects of such decision-making: decisions directly affecting industrial location, for instance, may induce a series of indirect effects on the location of residential, commercial, agricultural, transportational and recreational activities.

The system of human landscape organisation is thus based on the operation of a stimulus-response mechanism in which morphological structure is the essential response to a series of stimuli generated by decision-makers at different resolution levels and diffused throughout the morphological system by the intricate interdependences linking together each of the elements and their attributes (figure 10.1).

The efficient and effective functioning of the landscape system is dependent on the correct application of appropriate stimuli. Certainly, the many problems of congestion, scarcity, surplus, growth, stagnation, development or decline that appear to characterise so many aspects of the present organisation of human landscapes, are nothing less than the ultimate response to the application of inappropriate stimuli. To that extent, any faults in the spatial organisation of society largely lie, as Cassius might have suggested, 'not in our stars, but in ourselves' (*Julius Caesar,* I: ii). It would indeed be 'curiouser and curiouser' if no attempt were made in coming years to improve on locational and environmental decision-making practices since spatial mechanisms, while intricate and complex, are the net product of man's own ingenuity. It is all too easy to be pessimistic about the present, and it is equally all too easy to be optimistic about man's ability to learn from experience, and to create a better future, but 'to complain of the age we live in, to mistrust the present possessors of power, to lament the past, and to

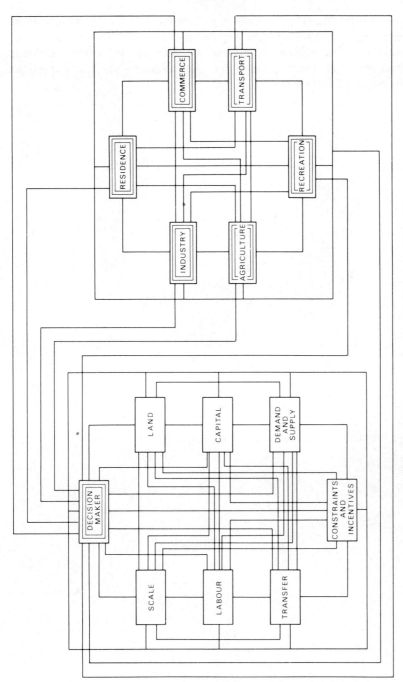

Figure 10.1 THE SYSTEM OF LANDSCAPE ORGANISATION.

conceive extravagant hopes of the future, are the common dispositions of the greatest part of mankind.' (Burke: *Thoughts on the Cause of the Present Discontent*). The analysis of organisation, location and spatial behaviour forms the vital starting point for the realisation of hopes, however extravagant, of the future utopia.

References

Journal title abbreviations follow the World List of Scientific Periodicals.
(UCDGRP–University of Chicago, Department of Geography, Research Papers.)

ABLER, R., ADAMS, J. S. and GOULD, P. (1971). *Spatial Organisation*, Prentice-Hall, Englewood Cliffs, N.J.
ACKERMAN, E. A. (1963). Where is the research frontier? *Ann. Ass. Am. Geogr.* **53**, 429–40.
ADORNO, T. W. et al. (1950). *The Authoritarian Personality*, Harper and Row, London.
ALEXANDER, J. W. et al. (1958). Freight rates: selected aspects of uniform and nodal regions. *Econ. Geogr.* **34**, 1–18.
ALLEN, G. R. (1954). Wheat farmers and falling prices. *Fm Economist*, **6**, 335–41.
ALLEN, S. W. (1959). *Conserving Natural Resources*, McGraw-Hill, New York.
ALONSO, W. (1965). *Location and land use*, Harvard University Press.
ARONSON, E. (1958). The need for achievement as measured by graphic expression. In J. W. Atkinson, *Motives in Fantasy, Action and Society*, Van Nostrand, New York.
ASHWORTH, W. (1954). *The Genesis of Modern British Town Planning*, Routledge and Kegan Paul, London.
AZZI, G. (1958). *Agricultural Ecology*, Constable, London.
BAER, W. (1964). Regional inequality and economic growth in Brazil. *Econ. Dev. Cultl Change*, **12**, 268–85.
BAIN, J. S. (1954). Economies of scale, concentration, and the condition of entry in twenty manufacturing industries. *Am. Econ. Rev.*, **44**, 15–39.
BARFØD, B. (1938). *Local Economic Effects of a Large Scale Industrial Undertaking*, Aarhus Oliefabrik, Economic research department.
BARTHOLOMEW, H. (1955). *Land Use in American Cities*, Harvard University Press.
BASKIN, C. W. (1966). *Central Places in Southern Germany* (translation of Christaller (1933), *Die Zentralen Orte in Süddentschland*), Prentice-Hall, Englewood Cliffs, N.J.

BAUMOL, W. J. and IDE, E. A. (1956). 'Variety in retailing', *Mgt Sci.*, **3**, 93–101.
BEAUJEU-GARNIER, J. (1966). *Geography of Population*, Longman, Harlow.
BECK, R. (1967). Spatial meaning and the properties of the environment. In D. Lowenthal, Environmental Perception and Behaviour, *UCDGRP*, **109**.
BERRY, B. J. L. (1959A). Ribbon developments in the urban business pattern. *Ann. Ass. Am. Geogr.*, **49**, 145–55.
BERRY, B. J. L. (1959B). The spatial organisation of business land uses. In W. Garrison *et al.*, *Studies of Highway Development and Geographic Change*, Washington University Press.
BERRY, B. J. L. (1967). *Geography of Market Centres and Retail Distribution*, Prentice-Hall, Englewood Cliffs, N.J.
BERRY, B. J. L., BARNUM, H. G. and TENNANT, R. J. (1962). Retail location and consumer behaviour. *Pap. Proc. Regl Sci. Ass.* **9**, 65–106.
BERRY, B. J. L. and GARRISON, W. L. (1958A). Functional bases of the central place hierarchy. *Econ. Geogr.*, **34**, 145–54.
BERRY, B. J. L. and GARRISON, W. L. (1958B). A note on central place theory and the range of a good. *Econ. Geogr.*, **34**, 304–11.
BERRY, B. J. L., SIMMONS, J. W. and TENNANT, R. J. (1963). Urban population densities: structure and change. *Geogrl Rev.*, **53**, 389–405.
BERRY, B. J. L., TENNANT, R. J., GARNER, B. J. and SIMMONS, J. W. (1963). Commercial structure and commercial blight. *UCDGRP*, **85**.
BERTALANFFY, L. Von. (1951). An outline of general systems theory. *British Journal of the Philosophy of Science*, **1**, 134–65.
BETHE, E. (1905). *Mythus-Sage-Märchen*, Leipzig.
BIBBY, J. S. and MACKNEY, D. (1969). *Land Use Capability Classification*, The Soil Survey, Technical monograph, **1**.
BLACKSELL, A. M. Y. (1971). Recreation and land use—a study in the Dartmoor National Park. In W. L. D. Ravenhill and K. J. Gregory, *Exeter Essays in Geography*, Exeter University, 1971.
BLAUG, M., PRESTON, M. and ZIDERMAN, A. (1967). *The Utilisation of Educated Manpower in Industry: A Preliminary Report*, Oliver and Boyd, Edinburgh.
BLOOM, G. F. and NORTHRUP, H. R. (1969). *Economics of Labour Relations*, Richard Irwin, London.
BLUMENFIELD, H. (1949). On the concentric circle theory of urban growth. *Land Econ.* **25**, 209–12.
BOGUE, D. J. (1949). *The Structure of the Metropolitan Community: A Study of Dominance and Subdominance*, University of Michigan Press.
BÖHNING, W. R. (1972). *The Migration of Workers in the U.K. and the European Community*, Oxford University Press.
BORTS, G. H. (1960). The equalisation of returns and regional economic growth. *Am. Econ. Rev.*, **50**, 319–47.
BORTS, G. H. and STEIN, J. L. (1964). *Economic Growth in a Free Market*, Columbia University Press.
BOUDEVILLE, J. R. (1966). *Problems of Regional Economic Planning*, Edinburgh University Press.
BRACEY, H. E. (1962). English central villages: identification, distribution and functions. *Land Stud. Geogr.* **24B**, 169–90.

BRANCHER, D. M. (1972). The minor road in Devon—a study of visitors' attitudes. *Reg. Stud.*, **6**, 49–58.
BRANDNER, L. and STRAUS, M. A. (1959). Congruence versus profitability in the diffusion of hybrid sorghum. *Rur. Sociol.*, **24**, 381–83.
BRECHLING, F. (1967). Trends and cycles in British regional unemployment. *Oxf. Econ. Pap.*, New Series, **19**, 1–21.
BRITISH RAILWAYS BOARD (1963). *The Reshaping of British Railways*, H.M.S.O.
BRITTON, J. H. (1967). *Regional Analysis and Economic Geography*, Bell, London.
BROOK, C. R. P. (1973). *Network structure in Ireland,* University of Exeter, Department of Geography, unpublished Ph.D. dissertation.
BROOKFIELD, H. C. (1969). On the environment as perceived, *Prog. Geogr.*, **1**, 51–80.
BROZEN, Y. (1950). *Implications of Technological Change*, Social Science Research Council, Chicago.
BRUSH, J. E. (1953). The hierarchy of central places in southwestern Wisconsin. *Geogr. Rev.*, **43**, 380–402.
BRUSH, J. E. and BRACEY, H. E. (1955). Rural service centres in southwestern Wisconsin and southern England. *Geogr. Rev.*, **45**, 559–69.
BUCKMAN, H. O. and BRADY, N. C. (1960). *The Nature and Property of Soils*, Macmillan, London.
BUNGE, W. (1962). *Theoretical Geography*, Gleerups.
BURTON, I. and KATES, R. W. (1965). *Readings in Resource Management and Conservation*, Chicago University Press.
BURTON, T. L. (1966). A day in the country. *Chart. Surv.*, **98**, 377–81.
BURTON, T. L. (1967). Outdoor recreation enterprises in problem rural areas. University of London, Wye College, Department of Economics, *Stud. Land Use*, Report 9.
BURTON, T. L. (1971). *Experiments in Recreation Research*, Allen and Unwin, London.
CAESAR, A. A. L. (1964). Planning and the geography of Great Britain. *Advt Sci.*, **21**, 230–34.
CAMERON, G. C. and CLARK, B. D. (1966). Industrial movement and the regional problem. University of Glasgow Social and Economic Studies, *Occasional papers No. 5*.
CARRUTHERS, W. I. (1957). A classification of service centres in England and Wales. *Geogrl J.*, **122**, 371–85.
CARRUTHERS, W. I. (1967). Major shopping centres in England and Wales, 1961. *Reg. Stud.*, **1**, 65–81.
CARTER, H. (1972). *The Study of Urban Geography*, Arnold, London.
CASSIRER, E. (1953). *The Philosophy of Symbolic Forms*, Yale University Press.
CHAPIN, F. S. (1965). *Urban Land Use Planning*, University of Illinois Press.
CHAPPELL, J. M. A. and WEBBER, M. J. (1970). Electrical analogues of spatial diffusion processes. *Reg. Stud.*, **4**, 25–39.
CHARLTON, W. (1970). *Aesthetics: An Introduction*, Hutchinson, London.
CHISHOLM, M. (1962). *Rural Settlement and Land Use*, Hutchinson, London.
CHISHOLM, M. (1963). Tendencies in agricultural specialisation and regional concentration of industry. *Pap. Proc. Regl Sci. Ass.*, **10**, 157–62.

CHISHOLM, M. (1966). *Geography and Economics*, Bell, London.
CHISHOLM, M. (1971). Freight transport costs, industrial location and regional development. In M. Chisholm and G. Manners, *Spatial Policy Problems of the British Economy*, Cambridge University Press.
CHORLEY, R. J. and KENNEDY, B. A. (1971). *Physical Geography: A Systems Approach*, Prentice-Hall, Englewood Cliffs, N.J.
CHRISTALLER, W. (1933). *Die Zentralen Orte in Süddentschland*, translated by C. W. Baskin (1966), Jena.
CIGNO, A. (1971). Economies of scale and industrial location. *Reg. Stud.*, **5**, 295–301.
CLARK, C. (1951). Urban population densities. *Jl R. Statist. Soc.*, Series A, **114**, 490–96.
CLAWSON, M., HELD, R. B. and STODDARD, C. H. (1960). *Land for the future*, Johns Hopkins, Maidenhead.
CLIFF, A. D. (1968). The neighbourhood effect in the diffusion of innovations. *Trans. Inst. Br. Geogr.*, **44**, 75–84.
COLE, J. (1960). The Kariba project. *Geography*, **45**, 98–105.
COLE, J. (1962). The Rhodesian economy in transition. *Geography*, **47**, 14–40.
CORTÉS, J. B. (1960). The achievement motive in the Spanish economy between the thirteenth and eighteenth centuries. *Econ. Dev. Cultl Change*, **9**, 144–63.
COSGROVE, I. and JACKSON, R. (1972). *The Geography of Recreation and Leisure*, Hutchinson, London.
COTTERILL, C. H. (1950). *Industrial Plant Location: Its Application to Zinc Smelting*, Webster Groves.
COWAN, P. and FINE, D. (1969). On the number of links in a system. *Reg. Stud.*, **3**, 235–42.
COX, K. R. (1967). Regional anomalies in the voting behaviour of the population of England and Wales, 1921–1951, University of Illinois, Department of Geography, unpublished Ph.D. dissertation.
COX, K. R. (1968). Suburbia and voting behaviour in the London Metropolitan area. *Ann. Ass. Am. Geogr.*, **58**, 111–27.
COX, K. R. (1969). The voting decision in a spatial context. *Proc. Geogr.*, **1**, 83–117.
COX, R. and ALDERSON, W. (1950). *Theory in Marketing*, Richard Irwin, London.
CRACKNELL, B. (1967). Accessibility to the countryside as a factor in planning for leisure. *Reg. Stud.*, **1**, 141–61.
CULLINGWORTH, J. B. (1970). *Town and country planning in England and Wales*, Allen and Unwin, London.
CULLINGWORTH, J. B. (1970). *Problems of an Urban Society*, 2 vols, Allen and Unwin, London.
CUMBERLAND, J. H. (1960). Interregional and regional input–output techniques. In W. Isard *et al. Methods of Regional Analysis*, M.I.T. Press.
CURRY, L. (1964). 'Landscape as system,' *Geogrl Rev.*, **54**, 121–24.
CURRY, L. (1966). 'Chance and Landscape.' In J. W. House (ed.), *Northern Geographical Essays*, Newcastle University, Newcastle-on-Tyne.
CURTIS, L. F., DOORNKAMP, J. C. and GREGORY, K. J. (1965). The description of relief in field studies of soils. *J. Soil Sci.*, **16**, 16–30.

DAVID, P. A. (1966). The mechanisation of reaping in the ante-bellum mid-west. In H. Rosovsky, *Industrialisation in Two Systems*, Wiley, New York.
DAVIES, R. L. (1968). Effects of consumer income on business provisions. *Urban Stud.*, **5**, 144–64.
DAVIES, W. K. D. (1966). The ranking of service centres: a critical review. *Trans. Inst. Br. Geogr.*, **40**, 51–65.
DAY, R. H. (1969). *Human Perception*, Wiley, New York.
DAWSON, J. A. (1973). 'Marketing.' In J. A. Dawson and J. G. Doornkamp, *Evaluating the Human Environment*. Arnold, London.
DAWSON, J. A. and DOORNKAMP, J. G. (1973). See previous entry.
DEWHURST, J. F. (1955). *America's Needs and Resources: A New Survey*, Twentieth Century Fund, New York.
DINSDALE, E. (1965). Spatial patterns of technological change. *Econ. Geogr.*, **41**, 252–74.
DOWNS, R. M. (1967). Approaches to, and problems in, the measurement of geographic space perception. University of Bristol, Department of Geography, Seminar Paper Series A, **9**.
DUNCAN, O. D. (1956). Optimum size of cities. In J. J. Spengler and O. D. Duncan, *Demographic Analysis: Selected Readings*, Free Press of Glencoe.
DUNN, A. S. (1954). *The Location of Agricultural Production*, Florida University Press.
DURKHEIM, E. (1954). *The Elementary Forms of Religious Life*, translation by J. W. Swain. Free Press of Glencoe.
ECKAUS, R. S. (1961). The North–South differential in Italian economic development. *J. econ. Hist.*, **21**, 285–317.
EDWARDS, W. and TVERSKY, A. (1967). *Decision Making*, Penguin, Harmondsworth.
EILON, S. *et al.* (1969). Analysis of a gravity demand model. *Reg. Stud.*, **3**, 115–22.
ESTALL, R. C. and BUCHANAN, R. O. (1966). *Industrial Activity and Economic Geography*, Hutchinson, London.
EVELY, R. and LITTLE, I. M. D. (1960). *Concentration in British Industry: An Empirical Study of the Structure of Industrial Production, 1935–51*, Cambridge University Press.
EYRE, S. R. (1964). Determinism and the ecological approach to geography. *Geography*, **49**, 369–76.
FINES, K. D. (1968). Landscape evaluation: a research project in East Sussex. *Reg. Stud.*, **2**, 41–55.
FISHER, J. L. and POTTER, N. (1964). *World Prospects for Natural Resources*, Johns Hopkins, Maidenhead.
FOUND, W. C. (1972). *A Theoretical Approach to Rural Land-Use Patterns*, Arnold, London.
FREEMAN, T. W. (1968). *Geography and Regional Administration*, Hutchinson, London.
FRIEDMAN, J. R. P. and ALONSO, W. (1964). *Regional Development and Planning*, Cambridge University Press.
FRIEDRICH, C. J. (1929). *Theory of the Location of Industries* (translation of Weber (1909), *Uber den standort der industrien*), Chicago University Press.

GALENSON, W. (1971). The employment problems of the less-developed countries: an introduction. In W. Galenson, *Essays on Employment*, International Labour Office, Geneva.

GARNER, B. J. (1960). Differential residential growth of incorporated municipalities in the Chicago suburban region (mimeographed).

GARNER, B. J. (1967). Models of urban geography and settlement location. In R. J. Chorley and P. Haggett, *Models in Geography*, Methuen, London.

GARRISON, W. et al. (1959). *Studies of Highway Development and Geographic Change*, Washington University Press.

GARRISON, W. L. and MARBLE, D. F. (1962). 'The structure of transportation networks,' U.S. Army Transportation Command, Technical Report, 62-111.

GEIGER, R. (1950). *The Climate Near the Ground*, Harvard University Press.

GETIS, A. and J. M. (1968). Retail store spatial affinities. *Urban Stud.*, **5**, 317-32.

GINSBURG, N. (1961). *Atlas of Economic Development*, Chicago University Press.

GLOAG, J. (1962). *Victorian Taste: Some Social Aspects of Architecture and Industrial Design*, Black, London.

GODDARD, J. B. (1970). Functional regions within a city centre: a study by factor analysis of taxi flows in central London. *Trans. Inst. Br. Geogr.*, **49**, 161-80.

GODLUND, S. (1956). Bus service in Sweden, *Lund Stud. Geogr.* (17B).

GOLANT, S. M. (1971). Adjustment process in a system: a behavioural model of human movement. *Geogrl Analys*, **3**, 203-20.

GOLDTHORPE, J. H. et al. (1968). *The Affluent Worker: Industrial Attitudes and Behaviour*, Cambridge University Press.

GOLLEDGE, R. G. and BROWN, R. A. (1967). Search, learning and the market decision process. *Geogr. Annlr*, **49**, 116-24.

GOLLEDGE, R. G., RUSHTON, G. and CLARK, W. A. V. (1966). Some spatial characteristics of Iowa's dispersed farm population and their implications for the grouping of central place functions. *Econ. Geogr.*, **42**, 261-72.

GOULD, P. R. (1966). On mental maps. Michigan Inter-University Community of Mathematical Geographers, Discussion paper 9.

GOODALL, B. (1970). Some effects of legislation on land values. *Reg. Stud.*, **4**, 11-23.

GOODE, W. J. (1951). *Religion Among the Primitives*, Free Press of Glencoe.

GORDON, I. R. (1973). The return of regional multipliers. *Reg. Stud.*, **7**, 257-62.

GOULD, P. R. (1969). Spatial diffusion. Commission on college geography, Resource paper 4, Association of American Geographers.

GOULD, P. R. and WHITE, R. (1968). The mental maps of British School Leavers. *Reg. Stud.*, **2**, 161-82.

GREEN, F. H. W. (1950). Urban hinterlands in England and Wales: an analysis of bus services. *Geogrl J.*, **96**, 64-81.

GREEN, H. L. (1955). Hinterland boundaries of New York City and Boston in southern New England. *Econ. Geogr.* **31**, 283-300.

GREENHUT, M. L. (1956). *Plant Location in Theory and Practice: The Economics of Space*, Carolina University Press.

GREGORY, K. J. and WALLING, D. E. (1973). *Fluvial Geomorphology: A Drainage Basin Approach*, Arnold, London.
GREYTAK, D. (1972). The firm in regional input-output analysis. *Reg. Stud.*, **6**, 327–29.
GRIME, E. K. and STARKIE, D. N. M. (1968). New jobs for old: an impact study of a new factory in Furness. *Reg. Stud.*, **2**, 57–67.
GROSS, B. M. (1966). *The State of the Nation*, Tavistock Publications.
GROTEWALD, A. (1959). von Thünen in retrospect. *Econ. Geogr.*, **35**, 346–55.
GUPTA, S. P. and HUTTON, J. P. (1968). *Economies of Scale in Local Government Services*, H.M.S.O.
GWILLIAM, K. M. (1970). The indirect effects of highway investment. *Reg. Stud.*, **4**, 167–76.
HÄGERSTRAND, T. (1952). The propagation of innovation waves. *Lund Stud. Geogr.*, **4B**, 3–19.
HÄGERSTRAND, T. (1953). *Innovationsförloppet ur Korologisk Synpunkt*, Gleerups; for translation (1967) see next entry.
HÄGERSTRAND, T. (1967). *Innovation Diffusion as a Spatial Process*, Chicago University Press; for Swedish original, see previous entry.
HAGGETT, P. (1965). *Locational Analysis in Human Geography*, Arnold, London.
HAGGETT, P. (1971). Leads and lags in inter-regional systems. In M. Chisholm and G. Manners, *Spatial Policy Problems of the British Economy*, Cambridge University Press.
HAGGETT, P. and CHORLEY, R. J. (1969). *Network Analysis in Geography*, Arnold, London.
HAGGETT, P. and GUNAWARDENA, K. A. (1964). Determination of population thresholds for settlement functions by the Reed–Muench method, *Prof. Geogr.*, **16**, 6–9.
HALL, E. T. (1966). *The Hidden Dimension*, Bodley Head, London.
HALL, P. (1965). *Land Values*, Sweet and Maxwell.
HALL, A. D. and FAGEN, R. E. (1956). Definition of system. *Yb. Gen. Syst.*, **1**, 18–28.
HAMILTON, F. E. I. (1967). Models of industrial location. In R. J. Chorley and P. Haggett, *Models in Geography*, Methuen, London.
HAMILTON, F. E. I. (1971). Decision-making and industrial location in Eastern Europe. *Trans. Inst. Br. Geogr.*, **52**, 77–94.
HANSEN, B. (1970). Excess demand, unemployment, vacancies and wages. *Q. Jl Econ.*, **84**, 1–23.
HARBISON, F. and MYERS, C. A. (1964). *Manpower and Education*, McGraw-Hill, New York.
HARRIS, C. D. and ULLMANN, E. L. (1945). The nature of cities. *Ann. Am. Acad. Polit. Soc. Sci.*, **242**, 7–17.
HARTSHORNE, R. (1939). *The Nature of Geography*, McNally.
HARVEY, D. W. (1963). Locational change in the Kentish hop industry and the analysis of land use patterns. *Trans. Inst. Br. Geogr.*, **33**, 123–44.
HARVEY, D. W. (1969). *Explanation in Geography*, Arnold, London.
HARVEY, D. W. (1973). *Social Justice and The City*, Arnold, London.
HAWLEY, A. H. (1950). *Human Ecology–A Theory of Community Structure*, Ronald, New York.

HEPWORTH, N. P. (1970). *The Finance of Local Government*, Allen and Unwin, London.
HERBERTSON, A. J. (1905). The natural regions of the world. *Geogrl Teach.*, **13**, 104–12.
HEWINGS, G. J. D. (1971). Regional input-output models in the UK. *Reg. Stud.*, **5**, 11–22.
HIMMELSTRAND, U. (1960). *Social Pressures, Attitudes and Democratic Processes*, Almqvist and Wiksell, Stockholm.
HIRSCH, W. Z. (1959). Inter-industry relations of a metropolitan area. *Rev. Econ. Stat.*, **41**, 360–69.
HMSO. (1972). *Agriculture in Scotland, Report for 1971* (Cmnd 4904). H.M.S.O.
HOOVER, E. M. (1948). *The Location of Economic Activity*, McGraw-Hill, New York.
HOYT, H. (1939). *The Structure and Growth of Residential Neighbourhoods in American Cities.* Federal Housing Administration, Division of Economics and Statistics, Washington D.C.
HUFF, D. L. (1960). A topographic model of consumer space preferences. *Pap. Proc. Regl Sci. Ass.*, **6**, 159–73.
HUGHES, R. B. (1961). Interregional income differences; self-perpetuation. *Sth. Econ. Jl*, **28**, 41–5.
HUMPHRYS, G. (1965). The journey to work in south Wales. *Trans. Inst. Br. Geogr.*, **46**, 85–96.
HYMAN, H. (1959). *Political Socialisation: A Study in the Psychology of Political Behavior*, Free Press of Glencoe.
INUKAI, I. (1971). Farm mechanisation, output and labour input: a case study in Thailand. In W. Galeson, *Essays on Employment*, International Labour Office, Geneva.
I.P.C. (1971). *Merger Policy*, Industrial Policy Group.
ISARD, W. (1956). *Location and Space Economy*, M.I.T. Press.
ISARD, W. et al. (1960). *Methods of Regional Analysis: An Introduction to Regional Science*, M.I.T. Press.
ISARD, W. and KUENNE, R. E. (1953). The impact of steel upon the Greater New York-Philadelphia Industrial Region: a study in agglomeration projection. *Rev. Econ. Stat.*, **35**, 289–301.
ISARD, W. and SCHOOLER, E. W. (1955). *Location Factors in the Petrochemical Industry*, University of Washington Press.
JACKSON, J. N. (1972). *The Urban Future*, Allen and Unwin, London.
JEFFREYS, M. (1954). *Mobility in the Labour Market*, Routledge and Kegan Paul, London.
JOHNS, E. M. (1965). *British Townscapes*, Arnold, London.
JOHNS, E. M. (1969). Symmetry and asymmetry in the urban scene. *Area*, **2**, 48–56.
JOHNSON, H. G. (1967). *Economic Policies towards Less Developed Countries*, Allen and Unwin, London.
JONES, E. (1966). *Towns and Cities*, Oxford University Press.
JONES, G. E. (1962). The diffusion of agricultural innovations. *Jl Agric. Econ.*, **15**, 387–405.
JONES, H. (1965). A study of rural migration in central Wales. *Trans. Inst. Br. Geogr.*, **46**, 31–45.

KAHN-FREUND, O. (1965). *The Law of Carriage by Inland Transport*, Stevens, London.
KANSKY, K. J. (1963). Structure of transportation networks. *UCDGRP*, **84**.
KATES, R. W. (1967). The perception of storm hazard on the shores of megalopolis. D. Lowenthal, *Environment Perception and Behavior*, *UCDGRP*, **109**.
KEEBLE, D. E. (1967). Models of economic development. In R. J. Chorley and P. Haggett, *Models in Geography*, Methuen, London.
KEEBLE, D. E. (1971). Employment mobility in Britain. In M. Chisholm and G. Manners, *Spatial Policy Problems of the British Economy*, Cambridge University Press.
KEMENY, J. G. and THOMPSON, G. L. (1957). Attitudes and game outcomes. In M. Dresher, A. N. Tucker and P. Wolfe, *Annals of Mathematical Studies*, Princeton University Press.
KENEN, P. B. (1964). *International Economics*, Prentice-Hall, Englewood Cliffs, N.J.
KERRY SMITH, V. and PATTON, A. R. (1971). Sub-market labour adjustment and economic impulses: a note on the Ohio experience. *Reg. Stud.*, **5**, 91–3.
KIMBLE, G. H. T. (1951). The inadequacy of the regional concept. In L. D. Stamp and S. W. Wooldridge, *London Essays in Geography*, London School of Economics and Political Science.
KINDLEBERGER, C. P. (1965). *Economic Development*, McGraw-Hill, New York.
KING, L. J. (1961). A multivariate analysis of the spacing of urban settlements in the United States. *Ann. Ass. Am. Geogr.*, **51**, 222–33.
KING, L., CASETTI, E. and JEFFREY, D. (1969). Economic impulses in a regional system of cities: a study of spatial interaction. *Reg. Stud.*, **3**, 213–18.
KLINGEBIEL, A. A. and MONTGOMERY, P. H. (1961). *Land Capability Classification*, US Department of Agriculture, Soil Conservation Service, Agricultural Handbook (210).
KLIR, J. and VALACH, M. (1967). *Cybernetic Modelling* translated by P. Doran, Iliffe, London.
KLUCKHOLN, C. (1956). Toward a comparison of value-emphases in different cultures. In L. White, *The State of the Social Sciences*, University of Chicago Press.
KNAPP, R. H. (1958). n-achievement and aesthetic preference. In J. W. Atkinson, *Motives in Fantasy, Action and Society*, Van Nostrand, New York.
KNIFFEN, F. (1951A). The American covered bridge. *Geogr. Annlr*, **41**, 114–23.
KNIFFEN, F. (1951B). The American agricultural fair. *Ann. Ass. Am. Geogr.*, **41**, 42–57.
KNIFFEN, F. (1965). Folk Housing–key to diffusion. *Ann. Ass. Am. Geogr.*, **55**, 549–77.
KNOSS, D. (1962). *Distribution of Land Values in Topeka, Kansas*, University Press of Kansas.
KUHN, A. (1966). *The Study of Society: A Multidisciplinary Approach*, Tavistock Publications.
KUNKEL, J. H. and BERRY, L. L. (1959). Retail image. *Jl Retailing*, **14**.

LANDSBERG, H. H. et al. (1963). *Resources in America's Future: Patterns of Requirements and Availabilities, 1960-2000*, John Hopkins, Maidenhead.
LASUEN, J. R. (1962). Regional income inequalities and the problems of growth in Spain. *Pap. Eur. Congr. Regl Sci. Ass.*, **8**, 169–91.
LAUBADÈRE, A., de. (1970). *Traité élémentaire de droit administratif*, Pichon and Durand-Auzias.
LAWTON, R. (1968). The journey to work in Britain: some trends and problems. *Reg. Stud.*, **2**, 27–40.
LAYARD, P. R. G. and SAIGAL, J. C. (1966). Educational and occupational characteristics of manpower: an international comparison. *Br. Jl Ind. Rel.*, **4**, 222–43.
LAYARD, P. R. G., SARGAN, J. L., AGER, M. E. and JONES, D. J. (1971). *Qualified Manpower and Economic Performance*, Allen Lane.
LEAN, W. and GOODALL, B. (1966). Aspects of urban land economics. *Estates Gaz.*
LEONTIEFF, W. (1966). *Input-output Economics*, Oxford University Press.
LESTER, R. A. (1964). *Economics of Labour*, Macmillan, London.
LEVER, W. F. (1972). The intra-urban movement of manufacturing. *Trans. Inst. Br. Geogr.*, **56**, 21–38.
LEVY, H. (1966). *Industrial Germany: A Study of its Monopoly Organisations and their Control by the States*, Frank Cass, London.
LEWIS, W. A. (1955). *Theory of Economic Growth*, Allen and Unwin, London.
LEWIS, W. A. (1966). *Development Planning*, Allen and Unwin, London.
LICHFIELD, N. (1971). Cost-benefit analysis in planning: a critique of the Roskill Commission. *Reg. Stud.* **5**, 157–83.
LIEBENSTEIN, H. (1957). *Economic Backwardness and Economic Growth*, Wiley, New York.
LIEPMANN, K. (1944). *The Journey to Work*, Oxford University Press.
LINGE, G. J. R. (1967). Government and the location of industry in Australia. *Econ. Geogr.*, **43**, 43–63.
LINTON, D. (1968). The assessment of scenery as a natural resource. *Scott. geogr. Mag.*, **84**, 219–38.
LIPSET, S. M. and BENDIX, R. (1964). *Social Mobility in Industrial Society*, University of California Press.
LOCKLIN, D. P. (1960). *Economics of Transportation*, Richard Irwin, London.
LOMAX, K. S. (1943). The relationship between expenditure per head and size of population of county boroughs in England and Wales. *Jl R. statist. Soc.*, **106**, 51–9.
LÖSCH, A. (1940). *Die räumliche Ordnung der Wirtschaft*, Jena; translated by W. H. Woglom (1954).
LÖVGREN, E. (1956). The geographical mobility of labour. *Geogr. Annlr*, **38**, 344–94.
LOWENSTEIN, L. K. (1963). The location of urban land uses. *Land Econ.*, **39**, 407–20.
LOWENTHAL, D. and PRINCE, H. C. (1965). English Landscape Tastes. *Geogr. Rev.*, **55**, 186–222.
LYNCH, K. (1960). *The Image of the City*, M.I.T. Press.

McBRIDE, G. A. (1970). Policy matters in investment decision-making. *Reg. Stud.* **4**, 241–53.
McCLELLAND, D. C. (1961). *The Achieving Society*, Van Nostrand, New York.
McCLELLAND, D. C., STURR, J. F., KNAPP, R. H. and WENDT, H. W. (1958). Obligations to self and society in the United States and Germany. *Jl abnorm. Psychol.*, **56**, 245–55.
McCLELLAND, W. G. (1966). *Costs and Competition in Retailing*, Macmillan, London.
McKINSEY & CO. (1971). *Defining the Role of Public Transport in a Changing Environment*, Stationery Office, Dublin.
McLOUGHLIN, J. B. (1969). *Urban and Regional Planning—A Systems Approach*, Faber, London.
MALTHUS, T. R. (1798). *First Essay on Population*, facsimile reprint, Macmillan, London (1966).
MANNERS, G. (1964). *The Geography of Energy*, Hutchinson, London.
MANSFIELD, E. (1961). Technical change and the rate of limitation. *Econometrica*, **24**, 741–66.
MANSFIELD, N. W. (1969). Recreational trip generation. *Jl Trans. Econ. Policy*, **3**, 149–58.
MANSFIELD, N. W. (1971). The estimation of benefits from recreation sites and the provision of a new recreation facility. *Reg. Stud.*, **5**, 55–69.
MARBLE, D. F. (1967). A theoretical explanation of individual travel behavior. Northwestern University, Department of Geography Research Paper 13.
MARSDEN, K. (1971). Progressive technologies for developing countries. In W. Galenson, *Essays on Employment*, International Labour Office, Geneva.
MARTIN, J. E. (1964). The industrial geography of Greater London. In R. Clayton, *The Geography of Greater London*, Philip, London.
MAUDSLEY, R. H. and BURN, E. H. (1970). *Land Law: Cases and Materials*, Butterworths, London.
MAY, K. O. (1954). Transitivity, utility and aggregation in preference patterns. *Econometrica*, **22**, 1–13.
MAYFIELD, R. C. (1963). The range of a central good in the Indian Punjab. *Ann. Ass. Am. Geogr.*, **53**, 38–49.
MEARS, L. and PEPELASIS, A. (1961). Determinants of economic development. In A. Pepelasis, L. Mears and I. Adelman, *Economic Development*, Harper and Row, New York.
MEDHURST, F. J. and PARRY-LEWIS, J. (1969). *Urban Decay*, Macmillan, London.
MIERNYK, W. (1955). Labour mobility and regional growth. *Econ. Geogr.*, **31**, 321–30.
MINISTRY OF LABOUR (1967). *Assistant with Industrial Training for Firms in Development Areas* (Pamphlet 66593), H.M.S.O.
MITCHELL, J. B. (1954). *Historical Geography*, English Universities Press.
MOORE, F. T. and PETERSEN, J. W. (1955). Regional analysis: an inter-industry model of Utah. *Rev. Econ. Stat.*, **37**, 368–83.
MORGAN, M. A. (1967). Hardware models in geography. In R. J. Chorley and P. Haggett, *Models in Geography*, Methuen.

MORGAN, W. B. and MOSS, R. P. (1965). Geography and ecology: the concept of the community and its relationship to environment. *Ann. Ass. Am. Geogr.*, **55**, 339–50.
MORGAN, W. B. and MUNTON, R. J. C. (1971). *Agricultural Geography*, Methuen, London.
MORRISON, W. I. (1973). Input-output analysis and the firm. *Reg. Stud.*, **7**, 253–56.
MOSES, L. N. and SCHOOLER, E. W. (1960). Regional cycle and multiplier analysis. In W. Isard *et al., Methods of Regional Analysis*, M.I.T. Press.
MOSS, R. P. (1963). Soils slopes and land use in part of S. W. Nigeria. *Trans. Inst. Br. Geogr.*, **32**, 143–68.
MOUNTJOY, A. B. (1963). *Industrialisation and Under-Developed Countries*, Hutchinson, London.
MURDIE, R. A. (1965). Cultural differences in consumer travel. *Econ. Geogr.*, **41**, 211–33.
MURPHY, R. E. and VANCE, J. E. (1954A). Delimiting the CBD. *Econ. Geogr.*, **30**, 189–222.
MURPHY, R. E. and VANCE, J. E. (1954B). A comparative study of nine central business districts. *Econ. Geogr.*, **30**, 301–6.
MURPHY, R. E., VANCE, J. E. and EPSTEIN, B. J. (1955). Internal structure of the Central Business District. *Econ. Geogr.*, **31**, 21–46.
MYINT, H. (1964). *The Economics of the Developing Countries*, Hutchinson, London.
MYRDAL, G. M. (1957). *Economic Theory and Underdeveloped Regions*, Duckworth, London.
NEWBY, P. T. (1971). Attitudes to a business environment: the case of the assisted areas of the South West. In W. L. D. Ravenhill and K. J. Gregory, *Exeter Essays in Geography*, Exeter University.
NORTHEY, J. F. and LEIGH, L. H. (1971). *Introduction to Company Law*, Butterworths, London.
NORTON, H. S. (1963). *Modern Transportation Economics*, Bobbs Merrill, London.
NURSKE, R. (1953). *Problems of Capital Formation in Underdeveloped Areas*, Oxford University Press.
NYSTUEN, J. D. (1967). A theory and simulation of intra-urban travel. In Northwestern University, Department of Geography, Research Paper 13.
NYSTUEN, J. D. and DACEY, M. F. (1961). A graph theory interpretation of nodal regions. *Pap. Proc. Regl Sci. Ass.*, 7, 29–42.
OLSSON, G. (1965). Distance and human interaction–a migration study. *Geogr. Annlr*, **47B**, 3–43.
O'RIORDAN, T. (1971A). Environmental management. *Prog. Geogr.*, **3**, 173–231.
O'RIORDAN, T. (1971B). Perspectives on resource management. Pion.
OSBORN, D. G. (1953). Geographical feautures of the automation of industry. *UCDGRP*, **30**.
OWENS, D. (1968). *Estimates of the Proportion of Space Occupied by Roads and Footpaths in Towns*, Ministry of Transport, Road Research Laboratory Report LR-74.
PARDOE, A. (1972). *A Practical Guide for Employer and Employee to the Industrial Relations Act 1971*, Jordan.

PARRY-LEWIS, J. (1971). Mis-used techniques in planning: the forecasts of Roskill. *Reg. Stud.*, **5**, 145–55.
PARRY-LEWIS, J. and TRAILL, A. (1968). The assessment of shopping potential and the demand for shops. *Tn Plann. Rev.*, **38**, 317–26.
PARSONS, T. (1951). *The Social System*, Tavistock Publications.
PARSONS, T. (1958). The sociology of religion. In *Essays in Sociological Theory*, Free Press of Glencoe.
PATERSON, J. H. (1972). *Land, Work and Resources*, Arnold, London.
PATMORE, J. A. (1973). Recreation. In J. A. Dawson and T. C. Downkamp, *Evaluating the Human Environment*, Arnold, London.
PEARCE, D. W. (1971). *Cost Benefit Analysis*, Macmillan, London.
PENNING-ROWSELL, E. C. and HARDY, D. I. (1973). Landscape evaluation and planning policy. *Reg. Stud.*, **7**, 153–60.
PERRY, P. J. (1969). Working-class isolation and mobility in rural Dorset, 1837–1936: study of marriage distances. *Trans. Inst. Br. Geogr.*, **46**, 121–41.
PHILBRICK, A. K. (1957). Principles of areal functional organisation in regional human geography. *Econ. Geogr.*, **33**, 299–336.
PIAGET, J. and INHELDER, B. (1956). *The Child's Conception of Space*, translation by F. J. Langdon and J. L. Lunzer, Routledge and Kegan Paul, London.
PILGRIM, B. (1969). Choice of house in a new town. *Reg. Stud.*, **3**, 325–30.
POOR, H. V. (1869). *Influence of the Railroads of the United States in the Creation of its Commerce and Wealth*, Journeyman Printers' Cooperative Association, New York.
POUNDS, N. J. G. (1959). *The Geography of Iron and Steel*, Hutchinson, London.
PRED, A. (1965). Industrialisation, initial advantage and American metropolitan growth. *Geogrl Rev.*, **55**, 158–85.
PRED, A. (1966). *The Spatial Dynamics of US Urban-Industrial Growth*, M.I.T. Press.
PRED, A. (1967). Behaviour and location. *Lund Stud. Geogr.*, **27B**.
PUTMAN, P. C. (1954). *Energy in the Future*, Macmillan, London.
RANKIN, D. C. (1973). Man-power planning. In J. A. Dawson and J. C. Doornkamp, *Evaluating the Human Environment*, Arnold, London.
RAPOPORT, A. (1957). Contributions to the theory of random and biased nets. *Bull. Math. Biophys.*, **19**, 257–77.
RATCLIFF, R. U. (1949). *Urban Land Economics*, McGraw-Hill, New York.
RICHARDSON, H. W. (1969). *Regional Economics*, Weidenfeld and Nicolson, London.
RIGGS, J. L. (1968). *Economic Decision Models for Engineers and Managers*, McGraw-Hill, New York.
ROBINSON, E. A. G. (1958). *The Structure of Competitive Industry*, Cambridge University Press.
ROBINSON, G. W. S. (1953). The geographical region–form and function. *Scott. geogr. Mag.*, **69**, 49–58.
ROONEY, J. (1967). The urban snow hazard in the United States: an appraisal of disruption. *Geogr. Rev.*, **57**, 538–59.
ROSKILL, Sir E. (1971). *Report of the Commission on the Third London Airport*, H.M.S.O.

ROSTOW, W. W. (1963). *The Stages of Economic Growth*, Cambridge University Press.
ROTHERNBERG, J. (1967). *Economic Evaluation of Urban Renewal*, Brookings Institution.
ROUND, J. I. (1972). Regional input-output models in the U.K.: a reappraisal of some techniques. *Reg. Stud.*, **6**, 1–9.
ROXBY, P. M. (1926). The theory of natural regions. *Geography*, **13**, 376–82.
RUSHTON, G. (1969). Analysis of spatial behaviour by revealed space preferences. *Ann. Ass. Am. Geogr.*, **59**, 391–400.
SAARINEN, T. F. (1966). Perception of drought hazard on the Great Plains. *UCDGRP*, **106**.
SAMUELSON, P. A. (1958). *Economics*, McGraw-Hill, New York.
SANT, M. E. C. (1967). Unemployment and industrial structure in Great Britain. *Reg. Stud.*, **1**, 83–91.
SCHULTZE, C. L. (1964). *National Income Analysis*, Prentice-Hall, Englewood Cliffs, N.J.
SEGAL, D. and MEYER, M. (1968). Levels in political orientation. In M. Dogan and S. Rokkan, *Quantitative Ecological Analysis in the Social Sciences*, M.I.T. Press.
SEYFRIED, W. R. (1963). The centrality of urban land values. *Land Econ.*, **39**, 275–85.
SHARP, C. (1965). *The Problem of Transport*, Pergamon, Oxford.
SILLITOE, K. (1969). *Planning for Leisure*, Government Social Survey Individual Studies Report SS388, H.M.S.O.
SIMMONS, I. G. (1966). Ecology and land use. *Trans. Inst. Br. Geogr.*, **38**, 59–72.
SIMMONS, I. G. (1973). Conservation. In J. A. Dawson and J. C. Doornkamp, *Evaluating the Human Environment*, Arnold, London.
SIMMONS, J. (1964). The changing pattern of retail location. *UCDGRP*, **92**.
SIMMONS, J. (1966). Toronto's changing retail complex. *UCDGRP*, **104**.
SIMMONS, J. (1968). Changing residence in the city. *Geogrl Rev.*, **58**, 622–51.
SIMON, H. A. (1957). *Models of Man*, Wiley, New York.
SLICHTER, S. H. (1961). *Economic Growth in the United States*, J. H. Furst.
SMAILES, A. E. (1946). The urban mesh of England and Wales. *Trans. Inst. Br. Geogr.*, **21**, 1–18.
SMAILES, A. E. (1971). Urban Systems. *Trans. Inst. Br. Geogr.*, **53**, 1–14.
SMAILES, A. E. and HARTLEY, G. (1961). Shopping centres in the Greater London area. *Trans. Inst. Br. Geogr.*, **29**, 201–13.
SMITH, R. D. P. (1970). The changing urban hierarchy in Wales. *Reg. Stud.*, **4**, 85–96.
SOPHER, D. E. (1967). *Geography of Religions*, Prentice-Hall, Englewood Cliffs, N.J.
SORAUF, M. (1965). *Political Science*, Prentice-Hall, Englewood Cliffs, N.J.
STAFFORD, H. A. (1963). The functional bases of small towns. *Econ. Geogr.*, **39**, 165–75.
STEFANIAK, N. J. (1963). A refinement of Haig's theory. *Land Econ.*, **40**, 428–33.

STEWART, G. A. (1968). *Land Evaluation*, Macmillan, London.
STODDART, D. R. (1963). Geography and the ecological approach. *Geography*, 50, 242–51.
STONE, P. A. (1964). The price of sites for residential building. *Prop. Devr.*
STONE, P. A. (1965). The prices of building sites in Britain. In P. Hall, *Land Values*, Sweet and Maxwell.
STONE, P. A. (1970). *Urban Development in Britain 1964–2004*, Cambridge University Press.
SUTHERLAND, A. (1969). *The Monopolies Commission in Action*, University of Cambridge, Department of Applied Economics, Occasional Papers, 21, Cambridge University Press.
SYMONS, L. (1967). *Agricultural Geography*, Bell, London.
TAAFFE, E. J., MORRILL, R. L. and GOULD, P. R. (1963). Transport expansion in underdeveloped countries: a comparative analysis, *Geogrl Rev.*, 53, 503–29.
TAAFFE, E. J. and GAUTHIER, H. L. (1973). *Geography of Transportation*, Prentice-Hall, Englewood Cliffs, N.J.
TATHAM, G. (1951). Environmentalism and possibilism. In G. Taylor, *Geography in the Twentieth Century*, Methuen, London.
TELLING, A. E. (1967). *Planning Law and Procedure*, Butterworths, London.
THOMAS, D. (1973). Urban land evaluation. In J. A. Dawson and J. C. Doornkamp, *Evaluating the Human Environment*, Arnold, London.
THOMAS, E. N. (1961). Toward an expanded central place model. *Geogrl Rev.*, 51, 400–11.
THOMLINSON, R. (1967). *Demographic Problems*, Dickinson, London.
THOMPSON, W. W. (1967). *Operations Research Techniques*, Bobbs Merrill, London.
THOMPSON, I. B. (1970). *Modern France*, Butterworths, London.
THORPE, D. and NADER, G. A. (1967). Customer movement and shopping centre structure. *Reg. Stud.*, 1, 173–91.
THORPE, D. and RHODES, T. C. (1966). The shopping centres of Tyneside. *Econ. Geogr.*, 42, 52–73.
THUNEN, J. H. von. (1826). *Der isolierte Staat in Beziehung auf Landwirtschaft und Nationalökonomie*, Fischer; translated by C. M. Wartenberg (1966).
TIETZE, C. (1963). The condom as a contraceptive. In *Advances in Sex Research*, Society for the scientific study of sex, Harper and Row, London.
TIETZE, C. and POTTER, R. G. (1962). Statistical evaluation of the rhythm method. *Am. J. Obstet. Gynec.*, 84, 692–98.
TINBERGEN, J. (1963). Quantitative adaption of education to accelerated growth. In H. S. Parnes, *Planning Education for Economic and Social Development*, O.E.C.D.
TOWNROE, P. M. (1969). Locational choice and the individual firm. *Reg. Stud.*, 3, 15–24.
TOWNROE, P. M. (1972). Some behavioural considerations in the industrial location decision. *Reg. Stud.*, 6, 261–72.
TOYNE, P. (1971). Customer trips to retail businesses in Exeter. In W. L. D.

Ravenhill and K. J. Gregory, *Exeter Essays in Geography*, Exeter University.
TOYNE, P. (1974). *Recreation and Environment*, Macmillan, London.
TOYNE, P. and NEWBY, P. T. (1971). *Techniques in Human Geography*, Macmillan, London.
TUAN, Y. F. (1968). Discrepancies between environmental attitudes. *Can. Geogr*, 12, 176–91.
ULLMANN, E. L. (1957). *American Commodity Flow: A Geographic Interpretation of Rail and Water Traffic based on Principles of Spatial Interchange*, University of Washington Press.
ULLMANN, E. L. (1968). Minimum requirements after a decade. *Econ. Geogr.*, 44, 364–69.
ULLMANN, E. L. and DACEY, M. F. (1962). The minimum requirements approach to the urban economic base. *Lund Stud. Geogr.*, 24, 121–43.
UNITED STATES NATIONAL RESEARCH COUNCIL (1969). *Resources and Man: A Study of Recommendation*, Freeman, London.
VITA-FINZI, C. (1969). Early man and environment. In R. U. Cooke and J. H. Johnson, *Trends in Geography*, Pergamon, Oxford.
WAGER, J. F. (1964). How common is the land? *New Society*, 30 July.
WALLER, R. A. (1970). Environmental quality, its measurement and control. *Reg. Stud.*, 4, 177–91.
WARTENBERG, C. M. (1966). *The Isolated State*, Pergamon, Oxford, translation of Thünen (1826) *Der isolierte Staat in Beziehung auf Landwirtschaft und Nationalökonomie*.
WEBER, M. (1904). *The Theory of Social and Economic Organisation*, translated by A. M. Henderson and T. Parson (1947), Oxford University Press.
WEBER, A. (1909). *Uber den Standort der Industrien*, Tübingen; translated by C. J. Friedrich (1929).
WEINBERG, M. A. (1971). *Take-overs and Mergers*, Sweet and Maxwell.
WESTERGAARD, J. (1957). Journeys to work in the London region. *Tn Plann. Rev.*, 28, 37–62.
WHITE, R. (1967). The measurement of spatial perception. University of Bristol, Department of Geography, Seminar Paper Series A 8.
WHITE, R. C. (1952). *These Will Go to College*, Western Reserve University Press.
WHITEHAND, J. W. R. (1972). Building cycles and the spatial pattern of urban growth. *Trans. Inst. Geogr.*, 56, 39–56.
WILLIAMSON, J. G. (1965). Regional inequality and the process of national development. *Econ. Dev. Cultl Change*, 13.
WILLIS, K. G. (1972). The influence of spatial structure and socio-economic factors on migration rates—a case study, Tyneside 1961-6. *Reg. Stud.*, 6, 69–82.
WINGO, L. (1961). *Transportation and Urban Land*, Johns Hopkins, Maidenhead.
WINTERBOTTOM, M. R. (1953). The relation of childhood training in independence to achievement motivation, University of Michigan, unpublished Ph.D. dissertation.
WOGLOM, W. H. (1954). *The Economics of Location*, Yale University Press; translation of Lösch (1940), *Die äumliche Ordnung der Wirtschaft*.

WOLF, C. and SUFRIN, S. (1955). *Capital Formation and Foreign Investment in Underdeveloped Areas,* Syracuse University Press.

WOLPERT, J. (1964). The decision process in spatial context. *Ann. Ass. Am. Geogr.,* **54,** 537–58.

WOOD, L. J. (1970). Perception studies in geography. *Trans. Inst. Br. Geogr.,* **50,** 129–42.

WRIGLEY, E. A. (1967). Demographic models and geography. In R. J. Chorley and P. Haggett, *Models in Geography,* Methuen, London.

YEATES, M. (1965). Some factors affecting the spatial distribution of Chicago land values 1910–1960. *Econ. Geogr.,* **41,** 57–70.

YOUNG, A. (1973). Rural land evaluation. In J. A. Dawson and J. C. Doornkamp, *Evaluating the Human Environment,* Arnold, London.

YUILL, R. S. (1965). A simulation study of barrier effects in spatial diffusion problems. Michigan Inter-University Community of Mathematical Geographers, Discussion Paper 5.

ZELINSKY, W. (1967). Classical town names in the United States: the historical geography of an American idea. *Geogrl Rev.,* **57,** 463–95.

ZIPF, G. K. (1949). *Human Behaviour and the Principle of Least Effort,* Cambridge University Press.

Index

Entries in bold type refer to illustrations

Abler, R., Adams. J. S., and Gould, P. 28
Absenteeism 124, **125**
Accessibility 67, 86, 105–7, 161, 171–2, 176, 179, 181
Achievement 40, 46–8, **48**, 74, 80, 81
Ackerman, E. A. 16
Adorno, T. W. 45
Aesthetics 40, 48–50
Age 28, 40, 45, 53, 135, 205
Agglomeration **53**, 57–8, 62–70, 83, 86, 91, 108–9, 123, 174–5, 182–3, 216
Agriculture 13, **14**, 39, 61, 64, 67, 71, 79, 92, 97, 112, 121, 127, 154, 156–9, 176, 181, 187, 191, 213–15, 237
Alexander, J. W., *et al.* 165
Allen, G. R. 189
Allen, S. W. 107
Alonso, W. 178
Amenity 88
Aronson, E. 49
Ashworth, W. 220
Azzi, G. 96, **97**

Baer, W. 70
Bain, J. S. 56, **57**

Barfød, B. 118
Bartholomew, H. 91
Baskin, C. W. 200
Baumol, W. J., and Ide, E. A. 203–4
Beaujeu-Garnier, J. 80
Beck, R. 49
Beliefs 40, 41–5, 47, 80
Berry, B. J. L. 204
Berry, B. J. L., Barnum, H. G., and Tennant, R. J. 208
Berry, B. J. L., and Garrison, W. L. 206, 207
Berry, B. J. L., Simmons, J. W., and Tennant, R. J. 84
Berry, B. J. L., Tennant, R. J., Garner, B. J., and Simmons, J. W. 86, 87, 177, **181**, 208
Bertalanffy, L. von 3
Bethe, E. 42
Bibby, J. S., and Mackney, D. 98, **99**
Bid Rents 175–82, **177**
Blacksell, A. M. Y. 78
Blaug, M., Preston, M., and Ziderman, A. 123, 124
Blight 86–9, **87**, **88**
Bloom, G. F., and Northrup, H. R. 118

Blumenfield, H.   183
Bohning, W. R.   233
Borts, G.   70
Borts, G., and Stein, J. L.   70
Boudeville, J. R.   70
Bracey, H. E.   207
Brancher, D. M.   78
Brandner, L., and Straus, M. A.   160
Brechling, F.   110
Britton, J. H.   142
Brook, C. R. P.   12
Brookfield, H. C.   53
Brozen, Y.   155
Brush, J. E.   207
Buckman, H. O., and Brady, N. C.   96
Bunge, W.   16
Burgess, E. W.   178, **179**
Burton, I., and Kates, R. W.   107
Burton, T. L.   75, 78, 107

Caesar, A. A. L.   245
Cameron, G. C., and Clark, B. D.   34, 172
Capital   14, **15**, 60, 89, 102, 109, 120, 127, 143–60, 184, 211, 234–41, 258
Carruthers, W. I.   9, 210
Cartels   215–16
Carter, H.   102, 208
Cassirer, E.   49
Central Business District   87, 102, 104–5, 181, 208
Central Place Theory   191–210, **195, 197, 198, 199, 200, 201**
Ceteris paribus   20, 201, 210
Chance   23, 53
Chapin, F. S.   43
Chappell, J. M. A., and Webber, M. J.   160
Charlton, W.   48
Chisholm, M.   70, 102, 129, 176, 183, 191, 243, 253
Chorley, R. J., and Kennedy, B. A.   4
Christaller, W.   198–201, **197, 200**

Cigno, A.   58
Clark, C.   84
Clawson, M., Held, R. N., and Stoddard, C. H.   107
Cliff, A. D.   20, 26
Climate   94–6, **95**
Cole, J.   160
Commerce   13, **16**, 71, 86, 89, 176,
Congestion   2, 86, 94, 182
Consumption   75–6, 145
Constraints and incentives   14, **15**, 109, 211–56
Contact frequency   20, **21**, 26, 135, 147–8, 157
Control system   4, 13, **15**, 258
Conveyancing   227–8
Cortes, J. B.   47
Cosgrove, I., and Jackson, R.   77
Cost–Benefit analysis   35
Cotterill, C. H.   173
Cowan, P., and Fine, D.   183
Cox, K. R.   44, 45
Cox, R., and Anderson, W.   204
Cracknell, B.   72, **72**, 107
Cullingworth, J. B.   212, 256
Cumberland, J. H.   115
Cumulative causation   67–70, **68, 69**, 108, 133, 211
Curry, L.   16
Curtis, L. F., Doornkamp, J. C., and Gregory, K. J.   96, **98**

David, P. A.   127
Davies, R. L.   210
Dawson, J. A.   210
Day, R. H.   28, 53
Decentralisation   67, 69–70, 134, 216–20
Decision-making   3, 13, **15**, 17–53, **19**, **53**, 54, 56, 73, 102, 106, 134–5, 139, 157–60, 161, 169, 172, 184, 211–12, 257
Demand and supply   14, **15**, 78–109, 109–18, 120, 131–3, 164, 184–210, **190, 192**, 250–6
Dewhurst, J. F.   151

Diffusion   18–26, 51, 82, 147, 148, 156–60, **157**
Dinsdale, E.   160
Disintegration   60–1
Distance   14, 103, 135, 137, 139, 161, 162–8, 206, **206**, 243–5
Divisibility   112, 159
Downs, R. M.   53
Duncan, O. D.   58
Dunn, A. S.   183
Durkheim, E.   42

Eckaus, R. S.   70
Economic development   2, 42, 73, **74**, 83, 146
Economies of scale   54–70, 83, 91, 108, 131, 169, 182, 184, 189
Education and social services   28, 40, 45, 47, 53, 68, 80–2, 120, 122–5, **123**, 135–8, **137**, 141, 144, 146, 152
Edwards, W., and Tversky, A.   53
Elasticity   214–18
Environment   1, 2, 28, 30, 47, 49, 73, 141, 152, 183
Estall, R. C., and Buchanan, R. O.   129
Evely, R., and Little, I. M. D.   215
Eyre, S. R.   16

Feedback   5–6, 15, 40, 53, 69–70, 211–12
Fines, K. D.   **101**, 102
Fisher, J. L., and Potter, N.   76
Found, W. C.   73
Freeman, T. W.   256
Friction of distance   161, 165, 172–82, 193
Friedman, J. R. P., and Alonso, W.   70

Galenson, W.   109
Garner, B. J.   7, 172, 180
Garrison, W., and Marble, D. F.   13
Geiger, R.   94
Getis, A., and J. M.   65
Ginsberg, N.   124

Gloag, J.   50
Goddard, J. B.   13
Godlund, S.   9
Golant, S. M.   16
Goldthorpe, J. H., *et al.*   127
Golledge, R. G., and Brown, R. A.   53
Golledge, R. G., Rushton, G., and Clark, W. A. V.   201
Goodall, B.   256
Goode, W. J.   42
Gordon, I. R.   142
Gould, P. R.   20, 40
Gould, P. R., and White, R.   40
Green, H. L.   9
Green, F. H. W.   9
Greenhut, M. L.   183
Gregory, K. J., and Walling, D. E.   7
Grime, E. K., and Starkie, D. N. M.   139
Gross, B. M.   43
Grotewald, A.   183
Growth   1, 71, 73–85, 109, 124, 143, 172
Gupta, S. P., and Hutton, J. P.   58
Gwilliam, K. M.   160

Hägerstrand, T.   18, 20, 26, 156
Haggett, P.   7, 9, 52, 142, 173, **175**, 196
Haggett, P., and Chorley, R. J.   12, 92
Haggett, P., and Gunawardena, K. R.   207
Hall, A. D., and Fagen, R. E.   3
Hall, E. T.   53, 78
Hall, P.   107
Hamilton, F. E. I.   53, 175
Hansen, B.   142
Harris, C. D., and Ullman, E. L.   **179**, 180
Hartshorne, R.   16
Harvey, D. W.   3, 5, 64, 67, 73, 212, 256
Hawley, A. H.   180

Hepworth, N. P.   58, 240
Herbertson, A. J.   3, 16
Hewings, G. J. D.   142
Hierarchy   7, **7**, 13, 197, 207–9, 211
Himmelstrand, J.   45
Hirsch, W. Z.   118
Hoover, E. M.   175, 183
Housing   71, 77, 86, 89, 91, 93–4, **93**, 96, 124, 152, 176, 178–80, 229
Hoyt, H.   **179**, 180
Huff, D. L.   28, 34, 205
Hughes, R. B.   69
Humphrys, G.   142
Hyman, H.   45

Imageability   32–4, **36**, 138
Income   28, 40, 53, **53**, 74–6, 135, 138–9, 144–7, **145**, 178, 205
Industrial relations   125–7
Industry   13, **14**, 34, 56–7, 61–70, 71, 79, 83, 86, 89, 92, 96, 102, 108, 112, 118, 121–2, 129, 143, 154, 172, 176, 187, 234, 237
Inflation   18, 235–6
Information   15, 17, 18–32, **22**, 44, 46, 51, 53, **53**, 134–8, 153, 257
Innovation   86, 121, 127, 155–60, 161, 212, 213, 243, 250–1
Input–output analysis   112–18
Inukai, I.   121, 155
Investment   125, 144, 150–60, 235
Isard, W.   58, 177, **200**, 204–5
Isard, W., and Kuenne, R. E.   118
Isard, W., and Schooler, E. W.   56
Isotropic surface   20, 176, 191, 194, 199, 205

Jackson, J. N.   16
Jeffreys, M.   140
Johns, E. M.   50
Johnson, H. G.   254
Jones, E.   84
Jones, G. E.   156, 158, 159
Jones, H.   142

Kahn-Freund, O.   245
Kansky, K. J.   183
Kates, R. W.   38, 39, 51
Keeble, D. E.   153, 220
Kemeny, J. G., and Thompson, G. L.   39, **40**
Kenen, P. B.   255
Kerry Smith, J., and Patton, A. R.   142
Kimble, G. H. T.   6
Kindleberger, C. P.   151
King, L. J.   208
King, L., Casetti, E., and Jeffrey, D.   142
Klingebiel, A., and Montgomery, A. D.   98
Klir, J., and Valach, M.   4
Kluckholn, C.   43, **43**
Knapp, R. H.   49
Kniffen, F.   156
Knoss, D.   104
Kuhn, A.   4
Kunkel, J. H., and Berry, L. L.   33, **33**

Labour   2, 14, **15**, 34, 57, 60, 108–42, 184, 211, 220, 230–3, 259
Land   14, **15**, 57, 59, 71–107, 143, 184, 211, 220–9, 258
Landsberg, H. H.   89, 91
Landscape   1, 3, 6–16, 18, 50, 98–102, **100**, **101**, 161, 257–9
Lapse rates   21, 85, 103, 135, 136, 204
Lasuen, J. R.   69, 154
Laubadere, A. de   227
Lawton, R.   134
Layard, P. R. G., and Saigal, J. C.   122
Layard, P. R. G., Sargan, J. L., Ager, M. E., and Jones, D. J.   123
Lean, W., and Goodall, B.   183
Leisure   76–9, **76**
Lester, R. A.   142
Lever, W. P.   183
Levy, H.   215

Lewis, W. A. 48, 146, 149, 160
Lichfield, N. 53
Liebenstien, H. 147
Liepmann, K. 120
Linge, C. J. R. 256
Linton, D. 100, **100**
Lipset, S. M., and Bendix, R. 47, 134
Location 28, 40, 52, 56, 57, 61, 67, 73, 86, 94, 105, 108, 120, 129, 133, 139, 141, 154–5, 161, 169, 172, 182, 187, 191–210, 219
Locklin, D. P. 162, 165, 247
Lomax, K. S. 58
Lösch, A. 198–201, **199**, **200**, 208, 210
Lövgren, E. 141
Lowenstein, L. K. 64
Lowenthal, D., and Prince, H. C. 50
Lynch, H. 53

Malthus, T. R. 89
Manners, G. 160
Mansfield, E. 156, 159, 160
Mansfield, N. W. 76, 107
Marble, D. F. 210
Marsden, K. 127
Maudsley, R. H., and Burn, E. H. 256
May, K. O. 35
Mayfield, R. C. 210
McBride, G. A. 160
McClelland, D. C. 46, 47, 81
McClelland, W. G. 141, 159
McClelland, D. C., Sturr, J. F., Knapp, R. H., and Wendt, H. W. 47
McKinsey & Co. 249
McLoughlin, J. B. 16
Mears, L., and Pepelasis, A. 151, 235
Medhurst, F. J., and Parry-Lewis, J. 87, **87**, **88**
Miernyk, W. 142
Minimax solution 39
Mitchell, J. B. 241

Mobility 56, 77–8, 102, 109, 120, 129, 133–42, 233
Monopolies 215–16
Moore, F. T., and Petersen, J. W. 110, 115, 117, 118
Morgan, M. A. 173
Morgan, W. B., and Moss, R. P. 16
Morgan, W. B., and Munton, R. J. C. 97
Morphological system 4, 13, **14**, 256–7
Morrison, W. I. 142
Moses, L. N., and Schooler, E. W. 117
Moss, R. P. 107
Motivation 15, 17, 40–53, **53**, 73, 134, 138, 178, 211, 257
Mountjoy, A. B. 146, 150
Movement minimisation 139, 172–5, 181, 182
Multiplier 110, 116–18
Murdie, R. A. 205
Murphy, R. E., and Vance, J. E. 105
Murphy, R. E., Vance, J. E., and Epstein, B. J. 65
Myint, H. 160
Myrdal, G. M. **68**, 69, **69**

Neighbourhood effect 20
Newby, P. T. 34, 172
Nodes and nodal structures 6–13, **11**, 44, 84, 115, 241
Northey, J. F., and Leigh, L. H. 235
Norton, H. S. 161, 162, 171
Nurske, R. 145
Nystuen, J. D. 210
Nystuen, J. D., and Dacey, M. F. 12

Obsolescence 85–9
Olsson, G. 135, 138, 139
Optimisation 50–3, 73, 94, 109
O'Riordan, T. 16, 107
Osborn, D. G. 91, 121
Owens, D. 92

Pardoe, A. 232
Parry-Lewis, J. 53
Parry-Lewis, J., and Traill, A. 210
Parsons, T. 42, 43, **43**
Paterson, J. H. 107
Patmore, J. A. 77
Pearce, D. W. 53
Perception 28–34, **29, 31, 32,** 39, 51, 98–100, 134–5, 150, 178, 202–4
Personality 28, 40, 45, 205
Penning-Rowsell, E. C., and Hardy, D. I. 102
Perry, P. J. 20
Philbrick, A. K. 7
Piaget, J., and Inhelder, B. 49
Pilgrim, B. 183
Planning law 220–7
Political values 40, 42, 44, 45
Poor, H. J. 247
Population density 20, 57, 84–5, **85,** 203–5
Population growth 79–84, **79, 81, 83,** 109, 147
Pred, A. 53, 68, **68, 69,** 70
Preferences 15, 17, 18, 32–40, **37, 41,** 48, 50, 53, **53,** 178, 205, 257
Probability 22, 23, 36–9, **37**
Production coefficients 113–15, **113, 114, 116**
Productivity 60, 89, 109–10, 112, 118–28, 131, 143, 145, 155
Putnam, P. C. 91

Random numbers 23–5
Rankin, D. C. 109, 142
Rapoport, A. 44
Ratcliff, R. U. 33, **36**
Recreation 13, **14,** 71, 96, 98, 102
Regional identification 6–13, **8, 10**
Resolution level 4, 6, 7, 13, 18, 40, 44, 51, 257
Retail activity 6, 30, 33, **36,** 57, 61, 65, 92, 108, 118, 129, 140, 181, **181,** 187, 258

Richardson, A. W. 150
Riggs, J. L. 53
Risk, 36, 40, 51, 152
Robinson, E. A. G. 60
Robinson, G. W. J. 6
Rooney, J. 49
Roskill, Sir E. 35
Rostow, W. W. 48, 146
Rothenberg, J. 107
Round, J. I. 142
Roxby, P. M. 4, 16
Rushton, G. 210

Saarinen, T. F. 39
Samuelson, P. A. 216
Satisfaction 40, 50–3, 73, 140, 150
Savings 143, 144–50, **145,** 235
Scale 14, **15,** 54–70, 83, 91, 108, 110–12, 121, 131, 146, 159, 171, 184, 212–20, 258
Scaling 34–6
Scarcity 1, 71, 109, 143
Schultze, C. L. 160
Search 17, 26–8, 51
Seyfried, W. R. 104
Sharp, C. 165, 169
Sillitoe, K. 77
Simmons, I. G. 107
Simmons, J. 142, 210
Simon, H. A. 51
Simulation 23–6, **24, 25**
Slichter, S. H. 118
Smailes, A. E. 16, 207
Smailes, A. E., and Hartley, G. 208
Smith, R. D. P. 210
Social class 45, 47, **48,** 159, 178, 205–6
Social values 40, 42, 44, 45
Soil 94–6, **98**
Sopher, D. E. 53
Sorauf, M. 45
Specialisation 59–60, 108, 121
Stafford, H. A. 210
Stefaniak, N. J. 183

Index

Stewart, G. A. 107
Stoddart, D. R. 16
Stone, P. A. 89, 91, 92, 102, 103
Sub-optimality 18, 27, 32, 51, **52**
Subsidies 155, 165, 167, 234, 236–41
Substitution 60, 89, 120, 127, 143, 146, 165, 170, 183, 220, 230
Sutherland, A. 216
Symons, L. 94, 97
System 3–7, **5**, **14**, **15**, 18, 40, 53, 53, 71, 73, 86, 89, 110, 127, 143, 146, 152, 161, 183, 184, 211, 235, **259**

Taafe, E. J., and Gauthier, H. L. 162, 183
Taafe, E. J., Morrill, R. C., and Gould, P. R. 241
Tariffs 252–6, **255**
Tatham, G. 16
Taxation 144, 155, 234–5, 251–2, **252**
Technology 2, 73, 75, 86, 91, 96, 109, 121, 127, 131, 144, 146, 151, 155–60, 161, 212, 246
Telling, A. E. 227
Thomas, E. N. 107, 208
Thomlinson, R. 107
Thompson, I. B. 215
Thompson, W. W. 53
Thorpe, D., and Nader, G. A. 210
Thorpe, D., and Rhodes, T. C. 208
Thunen, J. H. von 175, 182
Tietze, C. 82
Tietze, C., and Potter, R. G. 82
Tinbergen, J. 122
Topography 94, 96, 98–100
Townroe, P. M. 53
Toyne, P. 31, **32**, 35, **36**, 77
Toyne, P., and Newby, P. T. 28, **177**, 208

Transfer 14, **15**, 91, 109, 139, 161–83, 184, 201, 211, 241, 250, 258
Transport 13, **14**, 20, 26, 34, 71, 92, **93**, 161–71, 187, 200–1, 213, 240
Tuan, Y. F. 53

Ullman, E. L. 210
Ullman, E. L., and Dacey, M. F. 210
Uncertainty 36–40
Unemployment 3, 18, 57, 109, 110, 117, 120, 132
Urbanisation 68, 83–5, **83**, 86

Vita-Vinzi, C. 16

Wager, J. F. 78
Waller, R. A. 107
Weber, A. 172–4, 182
Weber, M. 42
Weinberg, M. A. 160
Westergaard, J. 142
White, R. 53
White, R. C. 47
Whitehead, J. W. R. 183
Williamson, J. G. 70
Willis, K. G. 139
Wingo, L. 183
Winterbottom, M. R. 46
Wolf, C., and Sufrin, S. 148
Wolpert, J. 51, **52**
Wood, L. J. 53
Wrigley, E. A. 80

Yeates, M. 104
Young, A. 107
Yuill, R. S. 20

Zelinsky, W. 156
Zipf, G. K. 139
Zoning, 177–82, **177**, 179